The New Role of the
Academies of Sciences in the Balkan Countries

NATO ASI Series

Advanced Science Institutes Series

A Series presenting the results of activities sponsored by the NATO Science Committee, which aims at the dissemination of advanced scientific and technological knowledge, with a view to strengthening links between scientific communities.

The Series is published by an international board of publishers in conjunction with the NATO Scientific Affairs Division

A Life Sciences	Plenum Publishing Corporation
B Physics	London and New York
C Mathematical and Physical Sciences	Kluwer Academic Publishers
D Behavioural and Social Sciences	Dordrecht, Boston and London
E Applied Sciences	
F Computer and Systems Sciences	Springer-Verlag
G Ecological Sciences	Berlin, Heidelberg, New York, London,
H Cell Biology	Paris and Tokyo
I Global Environmental Change	

PARTNERSHIP SUB-SERIES

1. Disarmament Technologies	Kluwer Academic Publishers
2. Environment	Springer-Verlag / Kluwer Academic Publishers
3. High Technology	Kluwer Academic Publishers
4. Science and Technology Policy	Kluwer Academic Publishers
5. Computer Networking	Kluwer Academic Publishers

The Partnership Sub-Series incorporates activities undertaken in collaboration with NATO's Cooperation Partners, the countries of the CIS and Central and Eastern Europe, in Priority Areas of concern to those countries.

NATO-PCO-DATA BASE

The electronic index to the NATO ASI Series provides full bibliographical references (with keywords and/or abstracts) to more than 50000 contributions from international scientists published in all sections of the NATO ASI Series.
Access to the NATO-PCO-DATA BASE is possible in two ways:

– via online FILE 128 (NATO-PCO-DATA BASE) hosted by ESRIN,
Via Galileo Galilei, I-00044 Frascati, Italy.

– via CD-ROM "NATO-PCO-DATA BASE" with user-friendly retrieval software in English, French and German (© WTV GmbH and DATAWARE Technologies Inc. 1989).

The CD-ROM can be ordered through any member of the Board of Publishers or through NATO-PCO, Overijse, Belgium.

Series 4: Science and Technology Policy – Vol. 16

The New Role of the Academies of Sciences in the Balkan Countries

edited by

Ch. Proukakis
University of Athens,
Athens, Greece

and

N. Katsaros
NCSR, "Demokritos",
Institute of Physical Chemistry,
Ag. Paraskevi Attikis, Greece

Kluwer Academic Publishers

Dordrecht / Boston / London

Published in cooperation with NATO Scientific Affairs Division

Proceedings of the NATO Advanced Research Workshop on
The New Role of the Academies of Sciences in the Balkan Countries
Athens, Greece
19–23 November 1996

A C.I.P. Catalogue record for this book is available from the Library of Congress

ISBN 0-7923-4773-0

Published by Kluwer Academic Publishers,
P.O. Box 17, 3300 AA Dordrecht, The Netherlands.

Sold and distributed in the U.S.A. and Canada
by Kluwer Academic Publishers,
101 Philip Drive, Norwell, MA 02061, U.S.A.

In all other countries, sold and distributed
by Kluwer Academic Publishers,
P.O. Box 322, 3300 AH Dordrecht, The Netherlands.

Printed on acid-free paper

TABLE OF CONTENTS

LIST OF PARTICIPANTS

Director

Prof. Dr. Charalambos PROUKAKIS, Professor of Medicine, University of Athens, Department of Medicine,Vice Rector of University of Athens, 30, Panepistimiou str., 106 79 Athens, Greece

Key Speakers

Albania

Prof. Dr. Petrit SKENDE, Academy of Sciences of Albania, "Fan Noli" Square, Tirana, Albania

Armenia

Prof. Fadey SARGSYAN, National Academy of Sciences of Armenia, 24 Bagramian Ave., 375019 Yerevan, Armenia

Prof. Yuri SHOUKOURIAN, National Academy of Sciences of Armenia, 24 Bagramian Av., 375019 Yerevan, Armenia

Prof. Yuri L. SARKISSYAN, State Engineering University of Armenia, 105 Terian Str., 375009, Yerevan, Republic of Armenia

Bulgaria

Prof. Dr. Nadejda PETROVA, Bulgarian Academy of Sciences, 15 Noemvri str. N1, BG-1040 Sofia, Bulgaria

Prof. Dr. Naum YAKIMOFF, Bulgarian Academy of Sciences, 15 Noemvri str. N1, BG-1040 Sofia, Bulgaria

Canada

Prof. Robert E. ARMIT, OCRI Technology Transfer Centre, Suite 400, 340 March Road, Kanata, Ontario, K2K 2E4, Canada

Prof. Boris P. STOICHEFF, The Royal Society of Canada, 225 Metcalfe Street, Suite 308, Ottawa, ON, K2P 1P9 Canada

Czech Republic

Prof. Václav PACES, Academy of Sciences of the Czech Republic, Národní 3, CZ-11142 Prague 1, Czech Republic

Denmark

Prof. Henning SØRENSEN, The Royal Danish Academy of Sciences and Letters, H.C. Andersens Boulevard 35, DK-1553 Copenhagen V

France

Prof. Marianne GRUNBERG-MANAGO, French Academy of Sciences, 23, Quai de Conti, 75006 Paris, France

Greece

Dr. Nikos KATSAROS, National Centre for Scientific Research "Demokritos", Institute of Physical Chemistry, 153 10 Ag. Paraskevi Attikis, POB 60228, Greece

Prof. Charalambos PROUKAKIS, University of Athens, 30 Panepistimiou str., 106 79 Athens, Greece

Prof. Dr. George PARISSAKIS, National Technical University of Athens, Department of Chemical Engineering, 9, Heroon Politechniou str., Polytechnioupolis Zografou, 157 73 Athens, Greece

Prof. Ioannis PESMATZOGLOU, Academy of Athens, 28 Panepistimiou str., 106 79 Athens, Greece

Moldova

Prof. Ion BOSTAN, Technical University of Moldova, 168, Stefan cel Mare Av., Kishinev 2004, Republic of Moldova

Prof. Valeriu DULGHERU, Technical University of Moldova, 168, Stefan cel Mare Av., Kishinev 2004, Republic of Moldova

Prof. Olga ILIASENCO, Academy of Sciences of Moldova, Modern Languages Department, 3 Academiei str. Kishinev, MD 2028 Moldova

Prof. Sergiu RADAUTSAN, Academy of Sciences of Moldova, Centre of Semiconductor Materials, 5 Academy Str., Kishinev, MD 2028, Republic of Moldova

Prof. Dr. Andrei VARTIC, "Basarabia" Cultural Foundation, 18, Sfatul Tsarii str., Kishinev 2012, Moldova

NATO

Dr. Alain H. JUBIER, NATO Scientific Affairs Division, B-1110 Bruxelles, Belgium

Netherlands

Prof. Dr. P.J.D. DRENTH, Royal Netherlands Academy of Arts and Sciences

Romania

Prof. Virgiliu CONSTANTINESCU, Romanian Academy, Calea Victoriei 125, 71102 Bucharest, Romania

Prof. Dr. Florin Teodor TANASESCU, Ministry of Research and Technology 21-25, Mendeleev, 70168 Bucharest, Romania

Russia

Acad. Konstantin V. FROLOV, Russian Academy of Sciences, Mechanical Engineering Research Institute (IMASH)

Prof. Efim MALITIKOV, International ZNANIE Association, 4 Serova Proezd, Moscow, 101813, Russian Federation

Prof. Yurij SOLODUKHIN, Moscow International Association "ZNANIE", 40, Frunzenskaya Naberejnaya ap. 15, Moscow Centre, Russia

Turkey

Prof. Dr. Ayhan ÇAVDAR, Turkish Academy of Sciences, Atatürk Bulvari No 221, Kavaklidere, Ankara - Turkey

Ukraine

Prof. Sergei ANDRONATI, A.V. Bogatsky Physico-Chemical Institute of the National Academy of Sciences of Ukraine, 86 Lyustdorfskaya doroga, Odessa 270080, Ukraine

United Kingdom

Prof. Malcolm JEEVES, Royal Society of Edinburgh, 22-24 George Street, Edinburgh, EH2 2PQ, United Kingdom

Participants

Prof. Dr. Emin RIZA, Academy of Sciences of Albania, "Fan Noli" Square, Tirana, ALBANIA

Prof. Dr. Aleko MINGA, Tirana University, Faculty of Natural Sciences, Bld "Deshmoret e Kombit", Tirana, ALBANIA

Ms Afroditi PATRONI, General Secretariat of Research and Technology, 14-18 Mesoghion str., 115 10 Athens, POB 14631, GREECE

Ms Penelope SPILIOTI, General Secretariat of Research and Technology, 14-18 Mesoghion str., 115 10 Athens, POB 14631, GREECE

Dr. Errikos FOCAS, General Secretariat of Research and Technology, 14-18 Mesoghion str., 115 10 Athens, POB 14631, GREECE

Dr. Micele BARONE, National Centre for Scientific Research "Demokritos", Institute of High Energy Physics, 153 10 Ag. Paraskevi Attikis, POB 60228, Greece

Dr. Neophytos PAPADOPOULOS, National Centre for Scientific Research "Demokritos", Institute of Nuclear Technology, 153 10 Ag. Paraskevi Attikis, POB 60228, Greece

Prof. Dr. Ayhan ULUBELEN, Turkish Academy of Sciences, Atatürk Bulvari No 221, Kavaklidere, Ankara, Turkey

Prof. Liudmila RADAUTSAN, Ministry of Economy, Dept. of Foreign Economic Relations, 1, Sq. Marii Adunari Nationale, Kishinev, MD 2012, Moldova

Dr. Nikos KONSTANDOPOULOS, General Secretariat of Research and Technology, 14-18 Mesoghion str., 115 10 Athens, POB 14631, Greece

Dr. Efstratios KARABATEAS, General Secretariat of Research and Technology, 14-18 Mesoghion str., 115 10 Athens, POB 14631, Greece

Dr. Panagiotis KOROGIANNAKIS, General Secretariat of Research and Technology, 14-18 Mesoghion str., 115 10 Athens, POB 14631, Greece

Prof. Nikos UZUNOGLU, National Technical University of Athens, Department of Electrical Engineering, 9, Heroon Politechniou str., Polytechnioupolis Zografou, 157 73 Athens, Greece

Ms Ruxandra RIMNICEANU, Ministry of Research and Technology, 21-25, Mendeleev, 70168 Bucharest, Romania

Ms Maria TANASESCU, Ministry of Research and Technology, 21-25, Mendeleev, 70168 Bucharest, Romania

Prof. Panagiotis SISKOS, University of Athens, Chemistry Dept., Zographos Campus, 157 73 Athens, Greece

Dr. Kate HATZI, General Secretariat of Research and Technology, 14-18 Mesoghion str., 115 10 Athens, POB 14631, Greece

PREFACE

It is well known that the Academies of Sciences in Western Europe have different goals than those of Eastern Europe mainly due to their independent status. Although some of the Academies in the West supervise research activities or some institutes, their main mission is to stimulate and access scientific developments in their own countries. In particular, they have a mission to advise their governments and other central bodies on science policy and organization of research.

The Academies of Central and Eastern Europe supervise numerous research institutes with a relatively large number of research scientists. Also, many of these institutes carry out basic and applied research isolated from that of universities and industry. Industry on a few occasions in the past sought solutions to its problems or the development of new products from the Institutes of the Academies of Sciences.

The challenges now facing the Academies of Central and Eastern Europe include the lack of adequate financing, the loss of status of scientific work, the defection of young researchers and the difficulties of recruiting new high level research staff. A major problem is the emerging lack of candidates for doctorate studies. The organization and financing of research in institutes, universities and industries and the role that the Academies of Sciences can play is also one of the subjects to be addressed. Public funding is limited and most of the funds available are directed towards applied research. Private funding is hardly available at this stage of transition to market economies and if there is, it is directed to applied research. The different strategies followed by other Academies of Sciences of Central and Eastern Europe in order to face these challenges are presented in this volume.

Although the same problems of transition exist in countries of Central and Eastern Europe, the solutions cannot be the same for all countries due to different long-standing traditions and recent history.

The Academies of Sciences of the cooperation partner countries in the Balkan area face the above-mentioned challenges which were presented during the workshop by prominent authorities. Also, the external and internal "brain-drain" of young and prominent scientists is another serious problem that Academies of Sciences in the Balkan countries must face.

Some cooperation partner countries maintained their research institutions under the supervision of the Academies while introducing new methods of management and eliminating research groups that did not demonstrate scientific efficiency (Hungary, Belarus, Poland, etc.).

In other countries (Latvia, Lithuania), the research institutes are no longer under the supervision of the Academies. Now, their main task is to advise the state on matters of science policy, promotion of scientific research and international cooperation.

The links between Academies and Universities was also discussed. How effectively links could be established not as consequences of administrative decisions but as a result of practicable measures remains a challenge for the near future.

The realization of this meeting with the support of the NATO Scientific Affairs Division gave the pursued results included in the present volume.

It was my pleasure to act as a Director of the NATO ARW and I am most grateful to the NATO Scientific Affairs Division for the financial support and especially to its programme director, Dr. Alain Jubier, whose contribution was essential so that this meeting would be effective. Twenty one communications were presented by distinguished scientists, representatives of the following countries: Albania, Armenia, Bulgaria, Canada, Czech Republic, Denmark, Moldova, NATO, Netherlands, Romania, Russia, Turkey, Ukraine and United Kingdom.

I am most grateful to my colleagues of the organizing committee: Dr. Nikos Katsaros, Greece, Prof. Y. Sarkissyan, Armenia, Prof. P. Skënde, Albania, Prof. S. Radautsan, Moldova and Prof. V. Constantinescu, Romania.

My special thanks to all participants without whose interest and contribution, the ARW would not have been possible.

Charalambos Proukakis
Vice Rector
University of Athens

THE NEW ROLE OF THE ACADEMIES OF SCIENCES IN THE BALKAN COUNTRIES

I. PESMAZOGLOU
Academy of Athens
28 Panepistimiou str, 106 79 Athens, Greece

It is with much interest that I attend the inaugural session of the International Symposium on "The New Role of the Academies of Sciences in the Balkan Countries". In my capacity as President of the Academy of Athens, I extend to all participants a warm welcome.

Modern developments in science and technology invite a new approach in promoting progress in knowledge and its applications. Advances in pure and applied sciences are becoming increasingly interdependent. The results obtained from various countries and research establishments should be systematically compared with respect to their coherence and implications. International cooperation is of growing significance in promoting advances in research and its applications. This is increasingly necessary between the Balkan countries, particularly within the scope of enlargement of the European Union.

Modern developments in three directions invite new efforts in collaboration and coordination in sciences and/or humanities:

First, scientific thought and experiment are linked to theoretical work both in the methods being applied and in their associations. Research in applied sciences and their results invite a deepening of the interdependence between scientific hypotheses, experimental work as well as in production or marketing.

Second, research and its applications gain increasing significance in modern life. International comparisons should be advanced in several directions with an overall view on economic progress.

Third, the international exchanges of information and analyses should be promoted at the national and international levels. I wish to stress the significance of what I would call the "inter-institutional and inter-disciplinary collaboration", which refers to exchanges of information and ideas between different disciplines or institutions.

We are well aware in Greece of these developments and necessities. In our dual geographical position and political orientation, as a Balkan country and a full and active member of the European Union, Greece steadily

Ch. Proukakis and N. Katsaros (eds.),
The New Role of the Academies of Sciences in the Balkan Countries, xvii–xix.
© *1997 Kluwer Academic Publishers. Printed in the Netherlands.*

supports the enlargement of the European Union to the Balkan as well as to the Central and Northeastern European countries.

It would be appropriate on this occasion to inform you of the character and scope of the Academy of Athens, especially of its attitudes with respect to the promotion of knowledge and of the general importance of mathematics.

The Academy of Athens has the special feature of being inspired by the conception and working operation of Plato's Academy, established in Athens in the 4th century BC as the Center of philosophic debates in search of truth, according to Plato's Theory of Ideas and the practice of presenting and discussing alternative or opposed "arguments and/or analyses". Plato's Academy as an institution designed to promote understanding and knowledge remained in operation for about 9 centuries and exercised wide influence in the ancient and hellenistic world. The creation of corresponding institutions was stimulated after the Renaissance and inspired by Enlightenment. In fact, most of the well known Academies in Europe were created in the 17th century. It is indicative of the importance and scope attached by the Greek people to the present Academy of Athens, that its revival and reactivation in the sciences, the humanities and the arts constituted one of the main objectives proposed to the National Assembly of the Greek people during the initial period of the 1821 Greek War of Independence. The outstanding Greek intellectuals and statesmen of the middle and later part of the 19th century, considered the reactivation of the Academy of Athens as a highly significant landmark of the rebirth of our national life and confirmation of continuity with our intellectual heritage.

My final remarks refer to the importance of m a t h e m a t i c s, as understood by Greek philosophers - especially those associated with the Academy of Athens.

Plato believed that knowledge of Mathematics constituted an essential condition for understanding and contributing to philosophy. He, therefore, thought that an inscription should be placed at the entrance of his Academy, according to which "No ignorant of Geometry should be admitted". And when Plato was asked what is the difference between a human being and an animal, the answer was "...that the human being can m e a s u r e...".

Of course, Plato's concentration on philosophy was induced by public life and p o l i t i c s, that is by understanding the motives and aims of action in society. Plato and his students believed that a State of Justice should be created, in which the dominant influence would emerge from "t h i n k i n g human beings" ...that is "the philosophers". He accordingly believed that life will improve when "Kings become p h i l o s o p h e r s or philosophers Kings". These principles appear to be widely accepted in the period that followed the initial Christian era. Plato's approach or intuition was neither refused nor denied during the Renaissance or by later well-known thinkers as

Descartes or Leibnitz, Pascal or Neuton, as well as by modern scientific thought. In the 20th century, mathematics have also fruitfully induced some significant analyses in theoretical and quantitative economics.

In the framework of these founding ideas, experiences and traditions, a Conference on "The New Role of the Academies of Sciences in the Balkan Countries" with representatives from Eastern and Western European countries evidently presents a great interest and could hopefully propose some constructive ideas.

I convey to you all a warm message of welcome and good wishes on behalf of the Academy of Athens.

THE UNIVERSITY RESEARCH SYSTEM IN GREECE AND ITS EFFECT ON SCIENCE POLICY

CH. PROUKAKIS
University of Athens
30, Panepistimiou str., 106 79 Athens, Greece

1. Introduction

According to the laws of Greek universities, the main goals of the universities are the following:

- Teaching, through which the necessary theoretical and practical knowledge is transferred to undergraduate and graduate students. Teachers of every grade participate in teaching, which is given by lectures, practical demonstrations, seminars, participation in experiments and other suitable teaching techniques.
- Research, whose main objective is the production of new knowledge. This production, however, needs the continuous follow-up of the relevant publications and presentations in meetings all over the world, resulting in the continuous upgradingof knowledge of the research staff.
- Social contribution, which is affected through different activities, such as the medical care of the local community, cultural activities such as excavations and expositions, contribution to the production of the international and bilateral relations of our country etc.

The academic staff of the university is classified in four categories of seniority, all of which have teaching and research duties but they differ in their right to participate in the administration.

Those four categories are the following in order of seniority:

a. Lecturers
The qualifications necessary for the selection of lecturers are the following:

- A doctor's degree (PhD or equivalent).
- Two original publications and two years of independent teaching, or two years of research work in a recognized research center or a combination of the above qualifications.

1

Ch. Proukakis and N. Katsaros (eds.),
The New Role of the Academies of Sciences in the Balkan Countries, 1–5.
© 1997 *Kluwer Academic Publishers. Printed in the Netherlands.*

- Equivalent professional work might be a qualification replacing No 2.
- Reasonable prospects that the candidate will follow-up successfully an academic career.

b. Assistant professors

The necessary qualifications for the selection of assistant professors are the following:

- Two years of independent teaching after the award of their doctor's degree, or two years of work in a recognized research center or equivalent professional work or a combination of the above qualifications.
- Original publications or a significant monograph.
- Reasonable prospects that the candidate will follow-up successfully an academic career.

c. Associate professors

The qualifications necessary for the selection of associate professors are the following:

- Four years of independent teaching after the award of their doctor's degree, or four years work in recognized research centers or equivalent professional work, or a combination of the above qualifications.
- Original publications, a number of which must be monographs.
- The research work of the candidate must have an international recognition.

d. Professors

The qualifications necessary for the selection of professors are the following:

- Six years of independent teaching after the award of their doctor's degree, or six years of work in recognized research centers or equivalent professional work, or a combination of the above qualifications.
- Original publications and monographs.
- Three years of teaching in post-graduate courses.
- International recognition of their contribution to original research, as shown by a high citation index, in journals with a high impact factor, invitation to give lectures in several universities or international meetings, organization of international meetings etc.

For the promotion of any one member of the staff to the next category of seniority, it is necessary that a large part of his total research work be carried out at the category he holds as a candidate. For example, to be promoted to associate professor, most of the research work of the candidate must have been carried out when he was an assistant professor and not a lecturer.

The above description of the qualification for the selection of the teaching and research staff shows clearly that the production of important research work is, according to the law, a key factor for the selection and promotion of the academic staff.

2. Funding of Research

The funding of research in Greece is completely inadequate. The contribution of industry and of the private sector, in general, is almost non-existing.

As expected, the main sources of research funding are different and include the following:

- The regular university budget, as determined by the ministers of economics and education.
- Grants from the European Union. This has been a major source of funding during the last years.
- Grants from the General Secretariat of Research and Technology. Most of the funds from this source are given to the Research Centers and only a small proportion is diverted to the universities.
- Several ministries support, with small amounts, applied research relative to their main interests, e.g. the Ministry of Health supports medical research, the Ministry of Agriculture, agricultural research, the Ministry of Civilization, research on cultural activities etc.

Private foundations, such as the Empirition Foundation and the Onassis public benefit foundation, award grants to post-graduate students, usually studying abroad, and advertise prizes for significant research work usually carried out in our country.

3. Administration of Research

The financial administration of research in the universities is decentralized. The researcher himself decides how he will spend his budget, according to his proposal and the approval he has received from the body that sponsors his research.

Specific accounts operate in all universities and research centers, which receive orders from the grant-holder and pay the requested amounts to research workers for the purchase of research equipment, for trips of researchers to participate in meetings etc. The grant-holder selects the research, technical and secretarial staff who will support his project, decides for their financial awards, and signs the necessary agreements. The overheads paid to the special accounts for their service might add up to 25% of the budget, but are usually lower. The special accounts pay some administration staff, but usually have a significant surplus, part of which is given to support further research projects.

4. Post-Graduate Studies

Post-graduate studies, during the last four years, have developed extensively based on the provisions of a law published in 1992. They are organized in two levels:

- Degrees equivalent to Master's, for which a study of one full year minimum is required, but in some cases five semesters is minimum. The maximum number of students in each course is determined in advance (numerus clausus) and the students are selected after a public advertisement in which the selection criteria are clearly stated. Also, the most senior academic body of our country, the Academy of Athens selects its new members with appropriate criteria, which include the research work of the candidates.

5. Cooperation with Research Centres

There is ample cooperation between universities and research centres. This cooperation is most obvious in the following fields:

- Joint appointments of staff. The law permits the appointment of members of the universities' research staff as research workers in the national research centers. Thus, an exchange of information, views and knowledge between universities and research centres becomes easier.
- Joint post-graduate courses and other scientific activities are organized between universities and research centres. Thus, a close cooperation develops.
- Common research grants are occasionally awarded to members of the staff of universities and research centres, resulting in close cooperation and common publications.

6. National Consultative Research Council

This Council proposes the research policy of the country. It consists of special sub-committees for different disciplines. The president of this council, as well as most of its members, is a university professor.

7. Scientific Councils of the National Research Centres

These councils supervise and evaluate the research work of the National Research Centres. The president and the members of those councils are usually university professors. There is one such council for each national research centre.

8. Conclusions

During the last years, it became more obvious that research is an activity that forms a significant part of university life. Research work is essential for the selection and promotion of the members of the academic staff. There are no members of the academic or the lower teaching staff who are not legally obliged to carry out research. The recent development and support of organized post-graduate studies is expected to provide to our country more young scientists and scholars suitably trained and properly motivated to devote much of their activity to research. The financial support provided by the European Commission to research projects is a key factor enabling many young people to engage in research and many senior people to devote much more of their time than before in planning, supervising and executing research.

Universities and their professors play a key role in training research workers, planning, proposing and executing research projects, advising the state on its research policies, obtaining research grants and influencing the national research policies on their goals and achievements.

In conclusion, the University Research System in Greece has a great influence on the Science Policy of our country.

THE ROLE OF THE FRENCH ACADEMY OF SCIENCES

MARIANNE GRUNBERG-MANAGO
French Academy of Sciences
23, Quai de Conti, 75006 Paris - France

1. Summary

The French *Académie des Sciences*, often defined as "Parliament of the learned world" is the forum for top-level exchanges between the best specialists in France, from every discipline in the sciences and technology.

The *Académie* has all along been determined to maintain in its heart the great cultural traditions, and indeed the great symbols which have made its reputation stand out and ensured its national prestige, by holding regular meetings open to the public, where the burning problems of science are discussed, and by welcoming scientists from the world over. It will be seen from this account that, in the course of time, it has become a top-level institution serving as reference and acting as Advisory counsel, not only to assist our decision-makers, teachers and heads of industries in their choices and defining directions to take, but also to respond to the growing curiosity of the public before the rapid development of the Sciences.

It also plays a key role in propagating the image of French science throughout the world, equally through its publications, symposia and the positions it takes on the social impact of research and technology.

2. A historical glimpse

Founded in 1666 by Colbert, the *Académie* long remained the privileged meeting place for the *élite* of a learned Society who came to debate on such areas as Astronomy, Botany, Mathematics, or indeed Philosophy. Members of this Royal Academy held their meetings in the Louvre, in the stately rooms they were accorded by Louis XIV, who did not disdain to make short intrusions to hear learned things being discussed. Abolished during the Revolution along with the other Academies (including the *Académie Française*, some 30 years its senior), to some extent "resuscitated" by Napoleon Bonaparte, himself a Corresponding member in the mechanics

Ch. Proukakis and N. Katsaros (eds.),
The New Role of the Academies of Sciences in the Balkan Countries, 7–13.
© 1997 *Kluwer Academic Publishers. Printed in the Netherlands.*

section (he took with him several "Academicians" during the Egypt campaign), the *Académie des Sciences* was placed with the other Academies under a federating institution: the future *Institut de France*, on the former site of the tour de Nesle facing the Louvre; on the Left Bank of the Seine, in the premises of the school the *Collège des quatre nations*.

Since 1832, the *Institut de France* includes 5 *académies* : from the oldest, *the Académie Française, Académie des Beaux-Arts, Académie des Inscriptions et Belles-Lettres,* to the latest *Académie des Sciences Morales et Politiques.*

The *Institut de France* offers common facilities : a beautiful library, rich with 1.500.000 volumes, meeting rooms and the most impressive "Dome" room, once the church of the *Collège des quatre nations.*

Every year, the 5 *Académies* hold a common meeting in the "Dome" room. It has also taken advantage of this diversity to organize colloquia on interdisciplinary subjects, like "creation and discovery", "science ethics and law", "genome patenting", or Aspects of ageing.

3. The French Academy of Sciences today

Since the reinstatement of the *Institut*, the French Academy of Sciences has not ceased to grow in size and importance. Gradually it has incorporated representatives from all the great areas of learning, at least in those termed the "exact" sciences : mathematics and the experimental sciences. The main disciplines represented are: Mathematics, Physics, Mechanics, Earth and Planetary Sciences, Chemistry, Molecular and Cell Biology, Zoology and Botany, Human Biology and Medical Sciences. Further to its vocation to assemble the best representatives of these disciplines from France, the *Académie* also includes several eminent foreign scientists as associate members.

Members and correspondents are organized into two "divisions", the first one including mathematics, physics and science of the Universe, the second dealing with chemistry and all fields of biology. Each division itself is parted into sections. There are 8 sections altogether.

At present, the *Académie des Sciences* has nearly 600 strong members (135), correspondents and foreign associate members, and constitutes the largest component of the *Institut de France*.

About ten years ago, desiring to be in closer, more direct liaison with advanced technology and industry and indeed to answer better society's great challenges, the Academic Council for Sciences Applications (CADAS) was set up. This is constituted by 175 full or associate members from the world of industry and professional associations. The *Académie des Sciences* and

CADAS work together continuously, in most instances involving the very large number of expert reports commissioned by the government.

The Academy is managed by a "Bureau" composed of a President, a Vice-President and two Perpetual Secretaries - one for each division, and by an Advisory Committee. The President and the Vice-President are elected for two years. Its delegation on international relations is in charge for the relations with the foreign academic world.

The *Académie* holds regular, weekly, working meetings where the burning questions of scientific and technological activity are aired through lectures and discussions. An up-to-date picture of the state of knowledge is drawn which can help foresee repercussions for socio-ethical and intellectual considerations (for education for example) as well as in economic terms.

4. The grand objectives

The Académie plays an important role in French scientific activities.

It is devoted to preserving the memory of the great names of science and their discoveries, maintaining research and the scientific culture of the country at the highest level, a presence on the international scene, and to promote links in the different areas of research and technology. It also acts as expert counsel for government bodies and industry alike.

4.1. CONVERSATION OF THE HERITAGE AND MEMORY OF THE WORLD OF LEARNING

The *Académie* has a wealth of manuscripts, correspondence and other written documents that present milestones in scientific thought along the centuries.

Consultation of the archives (a department recently refurbished) will bring to light moments such as the debates that raged around Louis Pasteur on spontaneous generation or on the beginnings of microbiology, the reports transmitted by Maupertuis from Lapland where he was sent by the *Académie* (with the blessing of Louis XV) to measure the length of an arc of a meridian-which led to the conclusion for a flattening of the Earth around the poles. Also to be found are Huyguens memoirs on calculation of probabilities and the physical nature of light, not to mention writings of scientists like Cuvier on palaeontology, the correspondence of d'Alembert, founder of mechanics, the notes of radioactivity discoverer Henri Becquerel, or of de Broglie, inventor of wave mechanics. This extremely rich legacy is indeed carefully preserved in the archives. But that is not all : it is used constantly as a source of illustrations and references for many different exhibitions and educational events, to the great benefit of children, students, researchers and teachers and to satisfy the public's interest.

4.2. MAINTAINING RESEARCH AND SCIENTIFIC CULTURE AT THE HIGHEST LEVEL

The Academy does not fund research but it helps research laboratories and researchers by awarding prizes and offering advice on certain high level scientific appointments. Since we have agreements with many *Académies* in the world, we also promote exchanges and visits between French and foreign scientists. In France, the *Académie* has numerous links with research organizations as INSERM, CNRS, and Ministries.

At the international level, the Academy is responsible for relations with scientific unions (through ICSU and the national committees) and organizations such as IAP (Interacademic Panel), ALLEA (All European Academies).

Since 1835, the *Académie* has published the scientific journal *Comptes rendus de l'Académie des Sciences*, which is short, original, reporting new discoveries, results and observations and aims to inform the international scientific community. They are divided into several volumes depending on the subjects. One volume is devoted to Mathematics, then there are Physics, Mechanics, Astronomy and Chemistry, Earth Sciences and, finally Life Sciences.

For a wider audience, including notably many top industrialists and the chiefs of a variety of institutions, a bimestrial newsletter *La lettre de l'Académie des Sciences et du CADAS* is published which relates the highlight events of the life of our company and its members. An information bureau to cater for the more general public and enlighten them on the most significant developments of research "in progress", and in a form easily accessible to all, is being created, in view of the Académie's concern that information on its activities should be available to and within reach of all.

It is, however, the rich debates enlivening the public meetings that give the great names of science the opportunity to confront their points of view. Six Nobel Prize winners count among its members, as do several holders of the Fields medal (equivalent of the Nobel Prize on Mathematics) and numerous laureates of prestigious international prizes. There are also several Nobel Prize winners among the foreign associate members.

In the course of public meetings or working committees or meetings with industrialists, all the great questions and problems of our time having any bearing on science or technology are tackled: energy, nuclear, industry, space environment, health, natural resources, new industrial manufacturing processes or advances in disciplines of mathematics, computing, telecommunications, biology and medicine.

In terms of popularization of science, the Academy supports, with others, "*Science Contact*", which helps the public and journalists to meet experts and obtain information on scientific results.

4.3. PRESENCE ON THE INTERNATIONAL SCENE

Through the Académie's foreign associate members, the science of some 40 countries is represented. In addition, many different kinds of links are maintained with the relevant institutions or academies throughout the world. Our company is, therefore, an integral part of a vast international network which assumes a multitude of tasks and objectives and thus facilitates the transfer of scientific information coming from the world research community to France or vice versa. The *Académie* therefore plays a key role for France as a nation, by its consultation and reflection helping prepare international exchanges and cooperation in the field of technology. The Ministry of Foreign Affairs, public research organizations and universities often call on it for advice.

The CARIST (Academic Committee for International Relations) promotes meetings where informal discussions and such reflections take place.

Particular attention is of course paid to strengthening scientific links with the European Union. The *Académie* played a major role in the creation of ALLEA (All European Academies) and participates eagerly in its activities. Through the activity of CADAS, the *Académie* participates also in exchanges involving engineers and representatives of the industrial world in Europe.

For some time, encouraged by its delegation on international relations, the *Académie* has initiated a new mechanism to welcome foreign researchers on working stays in France called the National Alfred Kastler Foundation (FnAK), after our sadly missed fellow member, the physicist Alfred Kastler, Nobel Prize winner and a great advocate of closer liaison between scientists from all over the world. This Foundation, which is highly successful, enjoys the support of our government authorities. As well as taking charge of young foreign researchers when they arrive in France, it continues to communicate regularly with them once they have returned to their country.

The *Académie des Sciences* is, moreover, well aware that it has a humanitarian and cultural role. A standing committee has just been constituted with the objective of encouraging our company to work more closely with developing countries on applied scientific topics which could be useful locally for the countries concerned. Two examples can be given. The *Académie* called recently a meeting on "Solar energy and Health" and is now studying the problems of mother and child in terms of health and nutrition. Another standing committee, CODHOS, aims to defend human rights of scientists throughout the world.

4.4. ACTING AS EXPERT ADVISER TO GOVERNMENTAL BODIES AND INDUSTRY

The very broad spectrum of learning and skills represented at the *Académie des Sciences*, its highly flexible organization into committees of experts, competent to analyze the most diverse scientific and technological questions, have not failed to attract the attention of the public authorities and the industrial sector. Everytime it was appropriate to analyse new prospects for applications or to anticipate the economic, social or ethical consequences of discoveries or new inventions.

This is why, for about 15 years, reviving a traditional function that supplied information to such widely different types of regime as the absolute monarchy, the Convention, the Empire or the republics constituted after the Second World War, the *Académie* is organized to intervene in a consultative capacity, whenever it is called upon to advise on the great issues that concern the government and society as a whole.

This is done by involvement of expert working parties, but also and above all, by conducting specialized studies and reports. These are undertaken either on the *Académie's* own initiative or on request from such authorities as Ministries, the National Assembly or the Senate. The resulting publications are distributed widely, not only to the bodies that commissioned them. Examples of topics covered by recent reports include : the greenhouse effect, chemical pollution, the effects on health of low dose ionizing radiation, the operation of nuclear power stations, transgenesis in agriculture, biodiversity, the effects of soft drugs, health effects of fuels and their combustion products. Other more general themes treated have been academic research, the future of various disciplines such as mechanics, toxicology, oceanography, the role of metrology, the development of biotechnology companies, the patentability of living forms and materials, gene therapy, evaluation of concepts involved in intellectual property.

These reports are, therefore, of interest, either directly for the main sectors of industrial (farming and food, transport, nuclear energy, pharmaceuticals) or the public sector decision-makers (in academic/university research, future of particular disciplines), or, more generally, society in its questioning on ethics and the repercussions of the advance of science.

5. Conclusion

This review of our activities shows that our Academy is still doing the tasks which were assigned more than 3 hundred years ago and reasserted to the *Institut de France*: that is "to foster research and creativity in the humanities, the sciences and the arts and to reflect as well to watch and to advise on behalf of the nation". The *Académie des Sciences* is the place where scientific debate takes place, and where the government may ask for expertise. The number of members and corresponding members has increased to reflect the growing diversity of scientific fields. And, more than ever, we feel the necessity to increase our international cooperation with the other *Académies* and worldwide organizations, in order to participate in debates on global issues.

Initially conceived and functioning as a meeting place for an elite, rather inward-looking, little by little the *Académie des Sciences* has changed, without losing its soul! Its life and work have been fully tuned to the realities of the contemporary world: to the benefit, we consider, of the Society in which it exists and also in the broader humanistic sense.

ACADEMIES IN TRANSITION

VIRGILIU NICULAE CONSTANTINESCU
Romanian Academy
Calea Victoriei 125, 71102 Bucharest, Romania

Part I. *The International Workshop on "Academies in Transition"*
(Sinaia, Romania 24 - 27 April, 1995)

The workshop was organized by the Romanian Academy in co-operation with the Department of European Integration of the Government of Romania and the UNESCO European Centre for Higher Education (CEPES) under the joint patronage of UNESCO and the Romanian National Commission for UNESCO, as well as that of the Commission of the European Union, and under the auspices of Mr Daniel Tarchys, Secretary General of the Council of Europe. Financial support was provided by the Romanian Government and, indirectly, by the European Union through the PHARE Programme, the UNESCO Office for Science and Technology in Europe (ROSTE), the Romanian Academy as well as by various Romanian sponsors (the Elias Foundation, the Class of Agricultural Sciences and Forestry of the Romanian Academy, the Cugir Factory Ltd.), and other local authorities.

The Organizing Committee was chaired by professor Virgiliu Nicolae Constantinescu, President of the Romanian Academy, and included representatives of the above-mentioned organizations that sponsored the meeting as well some European academic personalities, namely professor Paul Germain, Perpetual Secretary of the French Academy of Sciences and President of the Steering Committee of ALLEA (All European Academies), professor Domokos Kosary, President of the Hungarian Academy of Sciences and professor Leszek Kuznicki, President of the Polish Academy of Sciences.

Nineteen national Academies of the countries of central and eastern Europe were represented by their presidents, vice-presidents, secretaries general or other high ranking executives, namely **Albania** (Shaban Demiraj, president and Ylly Vejsiu, vice-president), **Armenia** (Gaguik Sarkissian, vice-president), **Belarus** (Leonid Suschenya, president and Galina Filimonova, counsellor to the president), **Bulgaria** (Jordan Malinowski, president), **Croatia** (Vladimir Stipetic, member of the Academy), **Czech Republic** (Václav Paces, vice-president and Petr Kratichvil, member of the

15

Ch. Proukakis and N. Katsaros (eds.),
The New Role of the Academies of Sciences in the Balkan Countries, 15–32.
© *1997 Kluwer Academic Publishers. Printed in the Netherlands.*

Council of the Academy), **Estonia**, (Jüri Engelbrecht, president), **Hungary** (Dénes Berény, president of the Debrecen Regional Centre of the Academy), **Latvia** (Talis Millers, president, Janis-Talivaldis Kristapsons, counsellor to the president), **Lithuania** (E. Vilkas, vice-president), **Macedonia** (Aleksandar Andreevski, vice-president and Tashko Georgievski, secretary-general), **Republic of Moldova** (Andrei Andries, president and Haralambie Corbu, vice-president), **Poland** (Leszek Kuznicki, president and Iacek Lornacki, director), **Romania** (Virgiliu Niculae Constantinescu, president, Dan Radulescu, vice-president, Aureliu Sandulescu, vice-president, and Ionel Haiduc, president of the Cluj-Napoca Branch of the Academy), **Russia** (Vladimir Koudriavtsev, vice-president), **Slovakia** (Stefan Luby, acting president and Ján Kazár, scientific secretary), **Ukraine** (Boris Evgenievich Paton, president and Anatolyi Petrovici Shpak, chief scientific secretary), **Yugoslavia** (Aleksandar Despic, president and Dusan Kanazir, member of the Academy).

In addition, the **French Academy of Sciences** and **ALLEA** were represented by professor Paul Germain, Perpetual Secretary of the Academy of Sciences and President of the ALLEA Steering Committee, the **Royal Swedish Academy of Sciences** by Dr Olof G. Tandberg, Foreign Secretary, the **German Max Plank Society** by professor Paul B. Baltes, Chairman of the Scientific Council while the **National Research Council of the USA** was represented by Dr Glenn Schweitzer, Director.

The **Academia Europaea** was represented by professor Hubert Curien, President and former Minister of Science and Research in France, while professor Raymond Daudel, President of the **European Academy of Arts, Sciences and Humanities** submitted a working paper. The President of the Ukrainian Academy also represented the **International Association of the Academies in Science for Foreign Co-operation**.

The international organizations were represented, for **UNESCO,** by Ms Carin Berg, director of **CEPES**, Mr Vladimir Kouzminov, Chief of **ROSTE**, and Mr Mircea Ifrim, Secretary General of the **Romanian National Commission for UNESCO**; for the **Commission of the European Union** by Ms Karen Fogg, Chief of the Delegation of the European Commission in Romania and by Mr Michele Genovese from DG XII; for the **European Science Foundation** by Mr Peter Colyer, co-ordinator of Scientific Work. Ms. Vera Boltho, head of Division of Higher Education and Research, the Council of Europe also submitted a paper.

Romanian National Institutions were represented by professor Liviu Maior, Minister of Education, professor Doru Dumitru Palade, Minister of Research and Technology, Mr Ghiorghi Prisacaru, State Secretary, Head of the Department of European Integration, Dr David Davidescu, president of the Class of Agricultural Sciences and Forestry of the Romanian Academy and Dr Adrian Toia, Director General, Ministry of Research and Technology.

Mr **Ion Iliescu, President of Romania**, participated actively in Session 2 and offered a lunch on April 25, 1995, while **Mr Nicolae Văcăroiu, Prime Minister of the Romanian Government** offered the closing reception on April 26, 1995, in which a number of Ambassadors of the participating countries also took part.

The total number of participants was 52.

A total number of forty working papers were presented within the workshop, while an additional six working papers were distributed to participants (authored by a few participants who, at the last moment, were unable to come to Sinaia). Finally, one last paper was prepared by the Romanian Academy containing excerpts from Francis Bacon's unfinished book "*New Atlantis*" which presents the author's view of the ideal Academy.

The workshop consisted of two full days of discussions, including an opening session, seven working sessions and a concluding summing up. The working sessions were devoted to the following:

Session 1. National Academies: Past and Future, chaired by professor Paul Germain, Perpetual Secretary of the French Academy of Sciences, and professor Virgiliu N. Constantinescu, President of the Romanian Academy. Six working papers were presented.

Session 2. Mission of National Academies, chaired by professor Stefan Luby, President of Slovac Academy of Sciences and by professor Talis Millers, President of the Latvian Academy of Sciences. Seven working papers were presented.

Session 3. Research within the Academy, chaired by professor Leszek Kuznicki, President of the Polish Academy of Sciences, and by professor Jüri Engelbrecht, President of the Estonian Academy of Sciences. Eight working papers were presented.

Session 4. Links Between Academy and University, chaired by professor Leonid Suschenya, President of National Academy of Sciences of Belarus and by professor Aleksandar Despic, President of the Serbian Academy of Sciences and Arts. Six working papers were presented.

Session 5. International Co-operation and European Integration: Assistance Programme for central and eastern European Countries, chaired by professor Hubert Curien, President of the Academia Europaea. Six working papers were presented.

Session 6. Training for Research, Scholarships and Brain Drain, chaired by professor Vladimir Nikolaevich Koudriavtsev, vice-president of the Russian Academy of Sciences, and by Paul B. Baltes, chairman of the Scientific Council, Max Plank Society. Five working papers were presented.

Session 7. The Charter of the Academy and Other Issues. The session was chaired by professor Andrei Andries, President of the Academy of Sciences of the Republic of Moldova, and by professor Shaban Demiraj, President of the Academy of Sciences of Albania.

The **Concluding** session was chaired by professor Virgiliu N. Constantinescu, President of the Romanian Academy, and by professor Jordan Malinowski, President of the Bulgarian Academy of Sciences.

Extra programme activities included a welcome dinner on April 24, a piano recital and a buffet dinner on April 25, a Chorus recital and a closing party on April 26, an excursion to the South Carpathian mountains followed by a farewell dinner on April 28. The conference took place in the Romanian mountain resort of Sinaia, former summer residence of the Romanian kings. Not only was the conference held in one of the palaces, but the participants were also able to visit the other palaces and the Sinaia Casino during the receptions and cultural activities.

The workshop offered a wealth of information on the present conditions and aims of the academies in both central and eastern Europe and in western Europe, thus providing an understanding of both similarities and differences. Discussions were open and informal, offering a valuable input for further reflection on the intricate subject of the organization of science and research. In spite of similar conditions of transition in the countries of central and eastern Europe, and particularly of almost identical difficulties, it appeared clearly that solutions cannot be the same for all countries, due to different long-standing traditions and to recent history.

The challenges now facing the academies of the central and eastern European countries include the lack of adequate financing, the loss of status of scientific work, the defection of young researchers and the difficulties in recruiting new ones. The lack of understanding of the importance of fundamental research from the political side was put forward. Private funding is hardly available at this stage of the transition to market economies, but even public funding is directed mainly to applied research.

Different strategies have been chosen to face these challenges. Belarus, Bulgaria, Estonia, Hungary, Moldova, Poland, Russia, and Ukraine aim at maintaining the research institutions under the academies, while introducing new managerial methods and eliminating research units that do not demonstrate scientific efficiency. Romania forms a particular case, because of its specific experience over the last twenty years, when the academy was more-or-less destroyed. Romania has chosen to reconstruct its academy according to its own tradition with a number of research institutions under the academy. Somewhat similar solutions are also existing in some western countries such as Sweden or the Netherlands.

Significant changes have been introduced into Latvia and Lithuania in which the academies have been reorganized as learned societies, having individual scientists as members. The research institutes are thus no longer under the supervision of the academies, which now instead have specific tasks such as monitoring the quality of research, elaborating science policy, and promoting scientific research as well as international co-operation. In

Estonia, where the academy has kept its research institutions, these have, however, established flexible links with the universities, taking into consideration the small size of the country. The Czech Republic has transformed its academy into a network of research institutions, thereby abolishing memberships for individual scientists. The Slovak Academy of Sciences has to a great extent chosen the same path as the Czech Republic; however, it has not abolished individual memberships.

The specific situation in the former German Democratic Republic after the reunification of Germany led to a restructuring of the institutes of research under the former GDR Academy guided by the needs of adapting research modalities to the western norms and of assuring quality, adequate financing and the promotion of young scientists on the basis of merits. This policy led to a reduction in the number of institutes and of their staff members. As several institutes were closed the total decrease of staff amounted to two thirds. However, a decision was reached to establish fifteen new Max Plank institutes before the year 2000, hiring staff through international competition, and observing the prevailing open recruitment procedures based on scientific merit.

New bodies have been set up in several central European countries, in particular for the funding of research on competitive grounds. In some of these countries, a dual funding system has been established. Salaries for basic staff in the research institutions and running costs for premises are allocated under a specific heading in the state budget, while additional funds for projects are made available on the basis of competition and an evaluation on scientific grounds by expert committees.

The harmful effects of brain-drain were repeatedly mentioned, but in the session devoted to this subject and elsewhere external brain drain appears to be less harmful than internal brain drain, as it keeps young scientists in science with the hope that at least some will one day return home, bringing a new and different experience with them. The problem of internal brain drain is more serious, as it implies that young people are definitely lost to science. However, the danger of both aspects of the brain drain are at this moment very serious, given its present magnitude: figures close to 40 per cent were cited for certain countries of the former USSR. The combined effect of brain drain and of the difficulties in recruiting, both to graduate studies in science and to research careers, results in a situation that some participants did not hesitate to qualify as disastrous. A somewhat more optimistic note was the proposal to replace the expression *brain drain* by *brain flow*, implying the possibility of a reverse flow.

The issue concerning the links between academies and universities was vividly discussed. The participants from the western countries, including the United States, underlined the importance of university research and close links between universities and the academies or even more. The academies, however, often have as their specific mission to be in charge of particular kinds of research that, because of its character, is not suitable for universities, such as being of very long term, of pronounced interdisciplinarity, or of being located abroad. These institutes work in close collaboration with one or several universities, as in the case of Germany.

Although most of the participants agreed on the necessity of creating and/or maintaining links with the universities, others warned of the consequences of dismantling an existing and well-functioning system that provided its value. Opinions also varied on how links could be established. A full range of measures were presented, from an integration of research institutes into universities, to joint projects, and the employment of institutional research staff to teach in universities. However, in order for such measures to be successful, they have to be initiated by the institutions themselves. They will fail if they are simply the consequences of administrative decisions. Any type of integration process is bound to require considerable time.

A closely connected question is that of post-graduate training, the *training for research*. The terminology used, however, seemed to be somewhat confused. The term *Ph.D.-training*, that was frequently cited, did not obviously always cover the mixture of disciplinary and methodological courses combined with research for the doctoral thesis that characterizes the American Ph.D. programme. In those European countries that have adopted this kind of doctoral training, the conferring of a doctoral diploma is the privilege of the universities, as it is the highest level of higher education. On the contrary, in some eastern European countries, doctoral degrees are also conferred by the research institutes under the academy. They admit university graduates to course programmes that are usually three years in duration, by means of competitive admission examinations.

A major problem in the countries of central and eastern Europe is the emerging lack of candidates for doctoral studies. There is at present little stimulus for students to enroll, as the prospects for employment are bleak and, in addition, the salaries in research institutes and universities are low. This situation, that has led to a loss of status of the scientific researcher and teacher and that further hinders recruitment to graduate studies, is an alarming danger signal to the whole system of higher education and research in regard to the future recruitment pool of staff.

The working session on *International Co-operation and European Integration: Assistance Programmes for Central and Eastern European Countries* gave an opportunity for the international organizations present to give an overview of their support activities for research, science and technology. The representatives of *UNESCO* described the UNITWIN scheme with the UNESCO chairs as a framework for promoting international co-operation in various domains of concern. The University-Industry-Science

Partnership Programme (UNISPAR) targets the methodological aspects of co-operation with industry. Special mention was made of the *ROSTE* project "*Transformation of the Scientific Communities in Europe*" on peer review procedures and research innovation management. The participants from the European Commission gave an overview of the different support programmes, essentially run by DG XII, such as the *fourth framework programme,* Co-operation Programme in Europe for Research on Nature and Industry through Co-ordinated University Study (COPERNICUS), and the Programme for Co-operation in Science and Technology with Central and Eastern European Countries (COST). It was underlined that these programmes should be seen as complements to the many bilateral initiatives already being taken. Finally the European Science Foundation was presented as a forum of national public bodies, responsible for science and research. (even though ESF itself is a non-governmental organisation) More than one body form a given country can adhere. The ESF works through networks, scientific programmes and topical conferences.

Of special interest were the new, recently established institutions for the promotion of cooperation among Academies and scientists. The *Academia Europaea*, created in 1988 with its secretariat in London, aims at being a forum for scientific research, promoting, among other things, co-operative efforts to put in place large equipment, such as that of the European Centre for Nuclear Research (CERN) that cannot be supported by one country alone. The *Academia Europaea* does not run research programmes, but encourages interdisciplinarity, organizes colloquia, supports young researchers, and publishes a review. The *Academia Europaea* is supported by several major companies and individual high level scientists.

ALLEA (All-European Academies) was created in 1994, with its secretariat presently located in France, as a network of academies of science, to stimulate debate on the functions of the academies, to promote the role of science in society, and to provide advice both to individual countries and to international organizations active in the field. Questions presently being dealt with concern fundamental research and the issue of patents.

The Ukrainian academy has created a network, the *International Association of the Academies of Science*, comprising the academies of the CIS countries and Vietnam, to promote the exchange of publications, the mutual publishing of scientific results, and joint projects.

The concluding remarks summarized the views expressed during the working sessions. The financial constraints were unanimously viewed as the biggest hurdle to overcome, as they have repercussions in all areas: retaining of and/or recruitment of staff; maintaining and promoting excellence; recruitment of graduate students; and provision of necessary equipment, scientific literature, and communication facilities.

By contrast, the solutions enacted or envisaged varied widely. Some countries advocate the establishment of strict quality assessment procedures to reduce the number of research institutes in order to concentrate on a few centres of excellence. Others stress the need for more efficient management procedures and more decentralization. The smaller countries promote close co-operation with the universities, some as far as merging the institutes with the universities.

Following the debates in the working sessions, the participants from both the western and the eastern parts of Europe had a better understanding of the reasons for these diversities. This knowledge should be particularly put to use by the international organizations so that they may better design co-operation and support programmes that would really benefit research in central and eastern Europe.

It was underlined that because this workshop was devoted to the academies, only a part of the total question of research could be covered. This really highlighted the need for an overall European discussion focusing on the organization and financing of research, addressing both the research institutes, the universities, and industry.

All the participants agreed that the workshop had been extremely useful, particularly because of its open and informal debates, and hoped that it would be followed by other similar events. The necessity of promoting science as a unique human undertaking for the benefit of societies was stressed. *The role of science must be made clear to governments, parliaments and public opinion.*

(Note: This part is basically taken from the report on the Workshop, prepared by Ms. Carin Berg, nominated Raporteur of the Workshop and published in *Higher Education in Europe*, vol. XX, No.4, 1995, edited by CEPES-UNESCO, Bucharest).

Part II. *The Romanian Academy*

The Romanian Academy was founded as a learned society in 1866 but on much older roots, practically in synchronism with western European countries, as a result of the phenomenon called the *"Renaissance"*. Thus, first attempts that led to what it is called now university are: the *"Latin School"* of Cotnari, Moldavia (1562) and the *"Jesuit College"* of Cluj-Napoca, Transylvania (1581); a learned society, the *"Academy of Targoviste"*, Valachia was founded in 1603 and comprised local members as well as learned men from France, Italy, Greece, etc. The insecurity which characterized the history of Romanians for many centuries did not allow these institutions (as well as some similar ones that followed) to last for more than a few dozen years. It was only after the first lasting attempt to unify two of the Romanian provinces that an Academy could have been built on a solid basis. However, this academy was founded together if not before the nation and the independent national state were established. As a consequence, the prestige of the Romanian Academy (as was the case with several other academies of this region of Europe) was particularly high in the 19th and the first half of the 20th century.

The first mission of the Academy was to prove the unity of the Romanian language and of the history of Romanians who were living at the confluence of three empires: the Austrian to the West, the Russian to the East and the Turkish to the South. Gradually, apart from humanities, science played a more and more important role. Correspondingly, titular and honorary members (both domestic and foreign) were active within the Academy, in a manner that bears some similarity with the Institute of France. The Academy increasingly supported and even carried on research in the first half of the present century. Thus, in his reception speech of 1923, a distinguished sociologist, Dimitrie Gusti (later on President of the Academy) - by starting from Francis Bacon's ideas - clearly advocated "*Academia Militans*", as opposed to the "*Passive Academy*".

Two attempts to dismantle the Academy were performed during the 45 years of communism after the Second World War. The first occurred in 1948 when a state Academy was forcefully installed. Over 100 outstanding members of the former Academy were not included into the new institution, especially in philosophy and history (most suffered typical persecutions); instead, a strong pressure to ideologize the institution took place. This situation was less dramatic in natural sciences. About five dozen outstanding scientists were accepted into the Academy, most of them relatively young and all having a strong western education. Knowingly and/or intuitively, they subtly put into operation D. Gusti's mentioned ideas of 1923 by gradually building up a number of scientific schools and research units. The main consequence of this endeavour was that fundamental science in Romania witnessed a real renaissance in mathematics, physics, chemistry, technology, medicine and biology, geology, etc. The peak of this development occurred around the year 1965. Paradoxically, Romanian scientists were known and respected world wide at that time, but almost disregarded in their own country and worse, accused almost in an open manner of lack of will to help the domestic industry and economy. This led to the second attempt - almost successful - to dismantle the Academy: all research units of the Academy were either dismantled or displaced to economic sectors and forced to perform only applied research. This policy hit not only the Academy - reduced to an almost vegetative state - but the entire Romanian fundamental research.

The reaction of Romanian society immediately after December 1989, namely to resuscitate the Academy and to put again under its umbrella some research units that had survived the mentioned dark times - is perfectly explicable. The period 1990-1994 witnesses the rebuilding of the Romanian Academy while the main concern today is to achieve a normal and efficient operation according to its traditions and to the present needs of Romanian society and intellectual community.

Thus, the General Assembly of May, 1994 specified the fundamental missions of the Romanian Academy as follows:

1. The traditional mission, associated to its prestige as being the highest scientific forum of the country and a pillar of the Nation, primarily through the value of its members. Thus, the Romanian Academy elects a number of titular and corresponding members selected from the most outstanding personalities in science, letters and arts, limited to a total of 181 (the population of Romania is of 23,000,000 people). Some 130 new members were elected after 1990 so that today the Romanian Academy comprises 175 titular and corresponding members. In addition, about 120 honorary members, foreign and Romanian, were also elected. One should note that the Romanian Academy keeps under the same umbrella science, arts and letters.

Some internal rules might also be worth mentioning. Thus, the number of corresponding members not exceeding the age of 65 years should amount to at least 30% of the total of titular and corresponding members. The president, the vice presidents and the secretary general cannot operate beyond the age of 75 years. Another internal rule is that a person cannot be considered for election while being a dignitary in the Romanian state system or a leader of a political party.

The Romanian Academy awards about 90 annual prizes for remarkable achievements in science, letters and arts.

The Romanian Academy possesses also a library of over 10 million volumes, periodicals, manuscripts and other documents as well as a Publishing House. A number of 92 different journals are published quarterly, as well as a monthly magazine, entitled "Academica".

The Romanian Academy has agreements with 39 Academies as well as with some national funding institutions, governmental and non-governmental international organisations.

2. The social/managerial mission, i.e., to perform competent and impartial studies and works on topics of national and international interest.

Thus, five projects are under development: *"Romania 2020"*, *"The Motivations of the Romanian Youth"*, a *"Treatise on Romanian History"*, the *"Complete Dictionary of the Romanian Language"*, an *"Encyclopaedia of the Romanian Literature"*.

Reports on specific scientific and economic issues are also currently performed mostly asked for by the Romanian Government.

The Romanian Academy was recently instrumental in a comprehensive endeavour to produce the strategy of integrating Romania into the European Union; this document was approved and signed by all parliamentarian political parties.

3. The active mission. In accordance to Act No. 4 of 5th January 1990, the Academy carries on research in its own units. Presently, the Academy has 65 units, comprising about 2,000 researchers (more than 60% have either a Ph.D. degree or are preparing a Ph.D. thesis); additional personnel amounts to about 50% of the previous figure. About half of the research effort is devoted to mathematics, natural sciences, economics and some social sciences, while half is of cultural research character, such as: language and literature, history and archaeology, philosophy and psychology, art history, etc.

The active mission of the Academy will be treated into more detail later on in this paper.

Ph.D. training is also performed within the research units of the Romanian Academy.

4. Another mission that goes back to the Founding Act of the *Romanian Literary Society* of 1 April 1866 (that became the *Romanian Academy* in 1879) is to be an **Academy of all Romanians**. Indeed, about 10 million Romanians are nowadays living outside Romania.

This mission is fulfilled in various ways, primarily by electing as foreign honorary members outstanding scientists and learned personalities of Romanian origin and by organizing special scientific and cultural meetings. Thus, the Romanian Academy together with the Romanian Cultural Foundation organized on 24-27 May, 1994, an International Conference on *Romania and Romanians in Contemporary Science*; over 300 scientists of Romanian origin were present together with about 50 foreign scientists who used Romanian scientific results and collaborated with Romanian scientists. Over 300 scientists residing in Romania were also present to this Conference. Several follow up projects are currently under way and another similar Conference will be organised in 1997.

The missions mentioned above, as well as several other topics related to "*Academies in Transition*" were discussed in April, 1995 within a three-day Workshop that gathered presidents, vice presidents and secretaries general of 19 academies of central and eastern Europe, representatives of a number of academies of western Europe and USA, European Academies, the European Science Foundation, the Council of Europe, the Commission of European Communities and UNESCO. A volume containing about 40 working papers and discussions of the Workshop is available in English, French and Russian (see the first part of this paper).

The Research Strategy. The R & D system in Romania is structured around four centres - the Ministry of Research and Technology, the Ministry of Education (through the National Higher Education Research Council), the Romanian Academy and a very incipient private sector. The functions and areas are split among these bodies which are supposed to interact and co-operate.

As already mentioned - by its founding Act (No. 4 of 5th January 1990) as well as by its Statutes - the Romanian Academy provides for the promotion of advanced fundamental research in every field of science and art. At this time of socio-economic reform and transition, when integration into Europe is Romania's strategic goal, a top priority is to develop the Romanian Academy's research network and system.

Systematic and comprehensive action requires a realistic and objective analysis of the present state and assessment of the current activities, as well as clearly stated objectives, measures and stages, and their relationships.

An articulated presentation of this analysis and strategy for development was recently achieved in our **"White Paper on the Romanian Academy Research Strategy"**. The present presentation is a condensed version of that document, approved by the General Assembly of the Romanian Academy, of November, 22nd, 1995. To summarize briefly, the four objectives that are the key to high-quality research can be stated as follows:

- Identifying the priorities (programmes, projects, themes) and concentrating the resources in order to get beyond the critical mass required by research of international relevance and to be beneficial to Romanian society.
- Permanently and systematically evaluating research at all stages - ex-ante, interim and ex-post - emphasising effectiveness and using feedback to stimulate high-performance research (and researchers).
- Reforming research financing in an effort to provide incentives and to promote efficiency, mainly competitive funding by programmes through grants.
- Developing closer co-operation between Academy research and university research for the benefit of both, with the stress falling on the formative character of research work.

The Present State of Romanian Academy Research. As mentioned earlier, there are 65 research units under the 14 Classes (speciality sections) of the Academy. The humanities and social sciences amount to 62%. Mathematics and chemistry are also well represented. There are no units in physics, only one in technical sciences and one for agricultural science; life sciences are under-represented. Two thirds of the units are based in Bucharest and the remaining operate by Academy subsidiaries in Iasi, Cluj-Napoca and

Timisoara. As to size - in terms of researcher numbers - small (45%) and middle-sized (40%) units prevail.

All units suffer from a shortage of assets in facilities and equipment which can hardly provide the conditions for research. Many of the facilities are old and badly worn. Documentation and information is poor by all accounts. This *de facto* picture contrasts sharply with the Romanian Academy's status as the lawful owner of a wide range of assets derived from the pre-war endowments and gifts which were nationalized by the communist regime and have yet to be recovered.

Fortunately, human resources are far better than material resources. From the total number of 4,461 employees, 2,562 (or 57% of the work force) are researchers - certified (1,990) or not (this latter group is comprised of young graduates undergoing training). Almost 20 per cent of the research workers are under 30 years of age, and some ten per cent are over 60: this raises the question of attracting young people to research. More than 800 researchers in the Academy (or 30% or so of the research staff) hold a Ph.D. degree and about as many are doctoral students.

Low salaries - in both absolute and relative terms (e.g., in respect of education) - and the lack of appropriate working conditions for high-quality research in certain fields largely account for the brain drain and brain waste through lack of logistic support for creative skills and talent.

Many of the current shortfalls are rooted in finance - in its size and structure - as well as in its allocation. Overall expenditure per researcher is far lower than in foreign countries (including ex-socialist countries) and below the all-country average. A good part of expenditure is made up of the core budget, salaries more particularly, leaving too little for research-proper facilities, supplies, documentation, mobility, etc.

A recent trend has been to increase competitive financing by programmes and especially by way of grants. The Romanian Academy Grants (GAR) initiated in the spring of 1995 have made a promising start that should be built on.

Research units enjoy scientific and management autonomy, with some 70 members of the Academy being on their management (including their Scientific Councils). As a rule, research programmes are proposed by units and submitted to the relevant Section, Presidium and eventually the General Assembly for approval (in a bottom-up approach). This ensures the much-needed scientific expertise. A decision was made for institute directorships to be nominated by search-committees on a competitive basis in an effort to raise the level of management: at mid-1995, this action is under way.

There was a strong need for the input-output evaluation of research activities. This action began in 1992 but only gained momentum in 1994: the first Evaluation Report on research institutes was issued early in 1995. A second much improved Report Update was approved by the General

Assembly of November 22nd, 1995 and is due for publication before the end of the year. Adding to that ex-post evaluation was the auditing of chemical institutes under a PHARE programme, which is intended to be extended to other fields in the near future, e.g., economics. Ex-ante evaluation of projects is needed for their selection for GAR, but also more generally for the selection of programmes (and themes) which are funded from the core budgets. So far, there has been little concern with this latter purpose, with the obvious result of too large a number of themes and a scattering of meagre human and material resources ("one person - one theme").

Publishing is the essential (but not the only) way of putting fundamental research to use. Scientometric methods to record and analyse statistically the published material and its impact (citations) are not yet used appropriately. Annually published materials average less than two papers per research worker. Output dissemination is hindered considerably by the shortage of equipment - desk-top publishing facilities and printing presses.

The total amount of papers published in peer review journals in 1994 was 1,700, with about 30% in international and/or foreign journals; about 80 monographs were published in 1994.

Research output is used in the economy on a contract basis - whenever such contracts can be secured - which account for an average 10 per cent of institutional revenue, but many occasionally account for as high as 25 per cent in certain cases (e.g., chemistry).

Formative activities assume various forms, but they are basically undertaken as part of the doctoral programmes offered by certain Academy institutes; also, when joint research work is conducted or assignments are held as a second job in higher education.

Directions and Objectives of the System's Reform. Foremost among the four key objectives is to select the priority programmes and promote excellence in research. In addition to the five priority programmes listed previously under the social/managerial mission of the Academy, programmes by field are being screened to allow a concentration of efforts and resources so as to ensure their competitiveness - nationally and internationally. A first objective is to revive the schools with a long tradition and a high prestige. New schools are contemplated (and expected) to emerge, particularly in the newly emerging interdisciplinary fields. However, there is the obvious need to provide a few major instruments for experimental research (of multiple use in order to render their utilisation more efficient). Last but not least, certain fields which have found in Romania particularly good conditions for development are to be encouraged. One such example is the discovery of the "Movila" cavern where unique forms of life have evolved intensively in spite of the particularly severe conditions such as lack of oxygen.

High-quality research can not be stimulated unless it is permanently evaluated at all levels. The chief way is the peer review - assisted by objective, quantitative, statistical criteria. The information system about R & D indicators, the measuring ones in particular, should be developed. Evaluation is to be regarded as a way to rationalize the system: directors are to be appointed by successful competition, institutes and teams are to be audited, funding will be targeted by increasing the share of competitive financing through programmes and grants, researchers are to be certified and re-certified.

A major aim is to reform financing generally, as a part of a complex financing system in which the share of core funding is declining whereas competitive allocation mechanisms become widespread on the basis of ex-post and/or ex-ante evaluation feedback. More efficient financing could be obtained by lump funding rather than presently used itemized system.

Integration, through as broad and comprehensive co-operation as possible, of Academy research with higher education research is to be achieved by joint research staff, by joint research projects and even by Academy-sponsored research centres established at universities. A Memorandum on this topic was signed during the summer of 1995 by the Ministry of Education and by the President of the Romanian Academy. Another important objective in support of formative activities is the expansion of doctoral and post-graduate programmes as well as pre-employment for senior-year undergraduates.

Collaboration as close as possible with the other two actors of the Romanian research structural system is sought, particularly with the Ministry of Research and Technology which supports not only applied research and technological development activities but also a number of fundamental research projects and institutions. A Memorandum, similar in some ways to that concluded with the Ministry of Education is under discussion.

Essential support for the expansion of research is to come through international co-operation. In addition to inter-academic agreements, participation in European research programmes is expected to grow. Bilateral co-operation - between institutes, teams and individuals - is to be widened.

With legal shortfalls hindering the expansion of research, providing an appropriate legal framework is important. These matters are dealt with at length in our Study. Proposals are made on the Research Law, the Researcher Statute (including incentives, promotion, wages, etc.), research financing, restoration of the Romanian Academy property to support its activities.

Stages in Reform Implementation. The first results are already apparent: the five programmes of national interest have been launched, a first Research Evaluation Report was issued, the Romanian Academy Grants for 1995 were launched, competitions have been under way for the positions of Directors of

Academy institutes, an International Workshop on "*Academies in Transition*" was organized in April, 1995 and unanimously considered "as an European scientific event", Academy institutes (of chemistry) were audited by an international team of experts, work has started on internal and international information network, etc.

Short-term measures provide for evaluations and self-evaluations and PHARE audits to continue and expand, GAR competitive financing to develop, the institutional network to be rationalized as obvious gaps (physics, advanced technology) are intended to be filled, lobbying for research legislation to be amended (including the own Academy Statute), a steep increase in the participation in European and international research programmes is expected.

Long-term measures will pursue such standards of excellence as are required for European integration. They concern science strategy through careful selection of research projects in an effort to revive Romanian schools of science of high standing and to establish new ones. Training will be aimed to adjust to the demands of the next millennium when information is to be heavily relied on, and when the standard of living, the quality of life and national life are expected to depend largely on intellectual skills and capabilities for cultural, scientific and technological innovative work through the effort of elites for the benefit of society.

The Academy and the Integration of Romania into the European Union. In pursuing its various missions, the policy of the Romanian Academy is to place itself above domestic politics. This attitude is deeply rooted into the tradition of the Institution. Indeed, one of the first decisions of the second General Assembly of August 16/28, 1868 was: "*By taking into account that far from having political goals or intentions, the Romanian Academic Society has the only goal to involve itself in cultivating and consolidating the Romanian language and to promote the development of letters and science among Romanians*".

However, in view of its social/managerial mission mentioned previously, the Romanian Academy commits itself to tolerance and democracy as goals of national interest and therefore to the aim of integrating Romania into the European Union and into Euro-Atlantic structures. This endeavour was first tackled within the project, entitled "*Romania 2020*". Thus, more than 15 meetings were organized in 1994 and 1995 under this heading - symposia, round tables, workshops - on related themes, such as: energy, agriculture, various fields of technology, health, communication, juridical problems, economy, scenarios of Romania's development, role of culture in future societies, research policy, education, etc. Within the same program outstanding personalities-representing international organizations, governmental and non-governmental bodies, such as the Club of Rome,

UNESCO, the World Academy of Arts and Sciences, the European Union, etc. - lectured in the Assembly Hall of the Romanian Academy on topics related to the future of Europe and of the World. Two most recent examples are Dr. Federico Mayor, Director General of UNESCO and Dr. Roman Herzog, the President of Germany. In addition, a TV serial under the aegis of the Romanian Academy, entitled *"Millennium"* is associated to this endeavour, as well as a number in programmes of Romanian universities.

One should also recall the already mentioned workshop related to *"Academies in Transition"*, organized by the Romanian Academy in April, 1995, which, among other things, contributed to a better understanding of some specific problems of the European regions and countries, and to a better harmonization of science policies in the Euro-Atlantic countries.

Finally, the contribution of the Romanian Academy to the strategy of integrating Romania into the European Union should be emphasised, both as an institution (through the already outlined research strategy and the ideas resulted from the various activities of the project *"Romania 2020"*) as well as through the contribution of over 40 members of the Romanian Academy who were particularly active in the National Commission which prepared the draft of this strategy. The Commission was headed by an outstanding economist, titular member of the Romanian Academy.

Two aspects at least are worth mentioning, namely a vision of the society for the next century and a project proposed for Romania, entitled *"Multilingua"*.

In exploring various scenarios for the post-industrial society of the 21st century and for the possible place Romania will have - some present tendencies were emphasised - particularly the effects of the Technological-Informational Revolution which leads to a lesser and lesser stratum of the society directly involved in the production of food and industrial products of current use. Instead, the majority of the population will be involved in the so called *"third sector"*, that of services and possibly more in the *"fourth sector"* dealing with the *"free time"* of the members of the society. Therefore, the future society has a great chance to become one with *"informational-cultural"* character.

The *"multilingua"* project started when discussing means of implementing in Romania the *"tolerant"* mentality that already proved to lead to most beneficial effects in western Europe. One simple but far reaching idea was to try to ensure the opportunity for every young Romanian to learn foreign languages. From here on, the project evolved to the present more comprehensive form, that is to aim to ensure that, in a reasonable number of years, every Romanian of the young age group should succeed to master the following four languages:

- the Romanian language, including the Romanian culture and history (as the specific Romanian contribution to the richness of the future multi-cultural Europe);
- two languages of international circulation;
- the language of the informatics;
- the artistic language.

The Romanian Academy is therefore committed to pursue any initiative which may contribute to a better future Europe.

THE NATIONAL RESEARCH CENTERS AND THEIR ROLE ON THE TECHNOLOGY TRANSFER AND SCIENCE POLICY OF GREECE

N. KATSAROS
NCSR "Demokritos", Inst. of Physical Chemistry
153 10 Ag.Paraskevi Attikis, Greece

1. Introduction

Science and technology are perceived as a key resource both for competitiveness and the long-term growth of national economics. Mastering new technologies is seen as a prerequisite for economic development. Increasingly research and technology generation and its further diffusion into the whole of the economic fabric are becoming higher priorities for governments world-wide. Developing or acquiring technology and implementing it successfully in industry are the stated aims of the corresponding policies in many countries.

The European Union is recognizing that research and development are essential components of economic growth and social development. They play an ever increasing role in improving the quality of life and strengthening the international competitiveness of the member-states. Therefore, the European Union's R & D policy is a basic component in structuring the European Single Market.

The rapid convergence of the Greek economy with the average level of the European Union demands, inter alia, the promotion of crucial areas of the economy where research and technological change act as catalysts, hence affecting growth and the role of the country in the international distribution of labour.

Moreover, there is a need for harmonizing the R&D activities of Greece with those of the European Union especially in keeping with the strategic priorities set in the Maastricht Treaty. The Maastricht Treaty gives R&D policy a new impetus: enhancing its industrial dimension while broadening its scope. R&D policy acquires a horizontal nature cutting across other major policies (environment, energy, transport, agriculture etc.).

Ch. Proukakis and N. Katsaros (eds.),
The New Role of the Academies of Sciences in the Balkan Countries, 33–43.
© *1997 Kluwer Academic Publishers. Printed in the Netherlands.*

2. Basic Characteristics of the Greek R&D System

The basic characteristics and weaknesses of the Greek R&D system in light of the above can be summarized as follows:

1. Gross Expenditure for R&D versus Gross Domestic Product GERD/GDP in Greece is limited to 0.47%. Since 1978 this ratio gradually increased from 0.2% to present levels which, however, are admittedly low compared with other countries in Europe and other industrialized nations (Table I). The average for the European Union is 2.1%, and for the developing countries in the Far East, South Korea has 2.5%, Taiwan 1.5% and Malaysia 1%.

TABLE I. The Relevant Percentages of Gross Expenditure for R&D versus Gross Domestic Product GERD/GDP for European and Industrialized Countries.

Country	GERD/GDP
Japan	2.98
USA	2.88
Germany	2.82
France	2.34
United Kingdom	2.27
Netherlands	2.16
Denmark	1.54
Italy	1.24
Belgium	0.86
Spain	0.81
Portugal	0.61

Experts argue that in order for R&D to have considerable impact on economic growth and social development, this indicator should be at least at the level of 1% of the GDP. Thus, a gradual annual increase is anticipated to that by the year 2000.

2. The participation of the productivity sector (public and private enterprises) in the Gross Expenditure for R&D remains minimal. Industrial R&D consumed over 26% of the GERD according to 1991 figures with an encouraging upward trend. For the USA and Japan this figure is approximately 70% and the average community level is about 65%. The respective figures for other countries are exhibited in Table II.

So, in Greece, the productivity sector represents 1/4 of the whole research activities while in the industrialized countries it exceeds the 1/2 and in some of them approaches 3/4. The involvement of the private sector with increased expenditure in research and development activities is necessary because this will also increase the demand for it. It will also increase technology transfer, connect research with the production and orient the Greek system of research and technology to the needs of the Greek economy.

TABLE II. Percentage of the Gross Expenditure for R&D (GERD) Consumed by Industry.

Country	Industrially consumed GERD
Germany	75%
Belgium	72%
Ireland	60%
Spain	56%
Switzerland	75%

The underlying objective of increasing the ratio GERD/GDP to the above mentioned 1% can be reached with increased R&D expenditures of the productivity sector from the current 26% to 40% with corresponding financing of the remaining 60% from the public sector.

3. Research in Universities and Research Centers is marginally connected with the production systems. One of the main reasons for it, is that only a very low percentage of the productive enterprises are interested in R&D activities, transfer of technology, innovation, utilization of research results, investments in high technology equipment and recruitment of R&D human capital.

4. The demand for R&D is very low. The limited participation of the productive sector in the GERD/GDP results in the low demand for R&D. Although it appears an encouraging trend, the total number of private enterprises involved in R&D activities increased from 114 in 1986 to 270 in 1991. Based on 1989 figures the flow of funds from enterprises to academic institutions for R&D services was very small excluding a few outstanding collaborative efforts of larger firms. However, 1992 figures demonstrate a changing trend in funds and a growing interest for collaboration with universities and research centers. The low demand for R&D may be attributed to the high level of dependence of the Greek productive system on ready-made and often out-dated technology acquired from the more technologically advanced countries. This technology is usually applied in an isolated, static way without corresponding efforts of assimilation and diffusion of technology that may lead to increased competitiveness of their products.

5. The research and technological organizations (universities, technological institutes, research centers) on the one hand and the productive sector on the other follow different paths responding to different priorities and orientations. Basic research is the only activity of many research groups while direct applications without any research remains the main goal of a large number of enterprises in the private sector.

6. The human capital engaged in R&D activities is also very small according to 1989 figures if accounted to 10,000 persons in Full Time Equivalent. Per 1000 employed we have 2.4 involved in research activities while Spain 3.8, Ireland 6.6, United Kingdom 10, Germany 14.3 France 11.9, Japan 13.8, USA 12.

TABLE III. Number of People Involved in Research per 1000 Employed

Country	Human Capital Engaged in R&D (per 1000 Employed)
Greece	2.4
Spain	3.8
Ireland	6.6
United Kingdom	10
France	11.9
Japan	13.8
USA	12
Germany	14.3

From the above mentioned figure of 10,000 people involved in research activities about 6,000 of them are researchers and the rest is supporting personnel. The number of researchers in universities (full-time equivalent) was estimated 3,000 in research centers, 2,000 and 1,000 in the productivity sector.

7. The majority of the human capital in the universities remains isolated and cannot be activated in industrial R&D projects due to various institutional and administrative constraints so typical to the Greek Academia. The dispersion of similar scientific units and equipment to various universities and the lack of collaboration among them inhibits the creation of the critical mass, necessary for effective and competitive action.

8. The Research Centers and Institutes have the appropriate institutional context for mobilizing the human capital towards applied industrial research programmes. These research organizations provide in many cases the necessary critical mass, the facilities and equipment as well as greater flexibility in rapidly adapting to scientific and technological change.

9. The R&D companies in several industrial sectors are quite significant constituents of the Greek R&D system. They were initiated by the General Secretariat of Research and Technology along the lines of similar intermediate organizations successful throughout Europe and operating under regulations governing private law companies in Greece. These companies are contributing to the effective transfer of know-how and technology to the productive units of the particular industrial sector focusing especially on the need of SME's. Six companies are operating so far covering sectors in metallurgy, ceramics, food and drinks, textiles/clothing/fibers, marine technology and leather goods. These companies are supervized either by the Universities, or the Ministry of Industry, Energy and Technology. It is noteworthy that efforts are made from the private sector to establish similar sectoral units for wood, paper, construction materials, polymers and composites etc.

10. The science and technology parks being developed in the four regions of Greece (Athens, Patras, Heraklion Crete and Thessaloniki) were new initiatives under the support of the structural funds of the European Union. These parks are expected to mobilize the local potential towards supporting high technology investments as well as in creating incentives for business activities in new technologies of local or international nature. The science and technology parks are based more on the concept of "supply push" of new technology rather than the above mentioned sectoral companies which focus on creating demand. With the institutionalization of S&T parks surrounding the research centers and neighboring university laboratories, it is expected that researchers and companies will come together for utilization and commercialization of research results and solution of problems in their enterprises. In Greece, due to lack of sufficient experience on activities related to S&T parks, a gradual approach has been considered more feasible in the form of establishing initially "incubators" for support of new high technology companies and eventually developing larger units.

11. Another characteristic of the Greek system of S&T is the overconcentration of R&D activities in the broader Athens area. It is considered that 65% of all R&D activities and 54% of the total money allocated were spent in the broader Athens area. Central Macedonia absorbed the 18%, Crete 9%, Western Greece 8% and all other areas of Greece absorbed the remaining 11%. In the broader Athens area, the R&D expenditure per 1000 people is $35,000, in Central Macedonia $22,000, in Western Greece $25,000, in Crete $22,000 and in the rest of the country less than $5,000. This unequal distribution of R&D activities and funds is due to

the State which tries to harmonize in accordance with the specific needs of the particular areas.

12. The impact of the EU programmes on R&D has significantly influenced the Greek system on S&T. Research contracts awarded to Greek teams amount on the average to over 3% of the available budget of the EU programmes. In keeping with the fair return assessment, this percentage well exceeds the national contribution to the Community Budget 1.3% and taking into consideration that Greece is having only 0.6% of the EU's human capital and 0.3% of R&D expenditures. It is moreover becoming clear that R&D carried out in Greece and financed from abroad, forms an important and continually increasing share of the country's total R&D activity. The proportion of the external financing in the GERD increased within a few years from 2% (1986) to 25% (1992) ranking Greece first among OECD countries as the level of external financing of the GERD. Undoubtedly, the research funds from the European Union account for the major portion of the external financing. However, the most significant impact of EU funds on the Greek R&D system has been the addition of a new qualitative dimension and an increased awareness and interest in the part of universities, research organizations, enterprises and government on the catalytic role of R&D in economic growth leading to a higher profile of R&D activities.

13. The structural funds of EU have a number of positive effects on the Greek R&D system which can be summarized as follows:

- created strong impetus for medium-term R & D policy-making based on realistic objectives;
- encouraged the shift of the research community towards more applied/industrial research programmes;
- fostered to some extent the collaborative links between R&D organizations and the production system;
- expanded and upgraded the R&D infrastructure;
- increased the R&D human capital and training activities of researchers;
- carried out high quality research in specific R&D areas;
- encouraged growing tendencies of enterprises to carry out R&D activities, technology transfer and innovation activities;
- stimulated part of the production system on the utilization and commercialization of research results;
- enhanced participation of Greece in EU competitive programmes.

3. Greek Universities and R&D Organizations

The Greek system of research and technology consists of the following organizations that are involved directly with the research activities:

3.1 UNIVERSITIES

In Greece, there exist 17 universities (Table IV) with approximately 6,000 faculty members. There are 11 technological institutes of higher learning geographically distributed around the country.

TABLE IV. Universities Existing in Greece

University	Faculty Members
National University of Athens	1724
University of Salonika	1872
University of Ioannina	351
University of Patras	487
University of Thrace	192
University of Crete	259
University of the Aegean	37
University of Thessalia	11
Ionion University	11
National Technical University of Athens	528
Agricultural University	136
Technical University of Crete	36
Economic University	97
Pantion University	136
University of Piraeus	76
University of Macedonia	40
School of Fine Arts	26

All universities and institutes of higher learning are supervised by the Ministry of Education and is contributing the 46% of GERD. The infrastructure of a number of universities is weak; they lack modern equipment to carry out high quality research and the interaction and collaboration with the productive system is far from been satisfactory. The community initiative with the acronym STRIDE brought closer the universities with the production system and developed the infrastructure in some universities.

3.2 RESEARCH CENTERS

The research centers and institutes, (Table V) that are supervised by the General Secretariat of Research and Technology that belongs to the Ministry of Industry, Energy and Technology, are 14 and have approximately 2,000 people in Full Time Equivalent.

TABLE V. Research Centers and Institutes in Greece

Research Center	Members
National Center for Scientific Research "Demokritos"	700
Institutes	
Informatics	
Material Science	
Microelectronics	
Molecular Biology	
Nuclear Physics and Radioisotopes	
Nuclear Technology	
Physical Chemistry	
Foundation of Research and Technology	430
Institutes	
Electronic Structure and Laser (Crete)	
High Temperatures Chemistry (Patras)	
Informatics (Crete)	
Mathematics (Crete)	
Mediterranean Studies	
Molecular Biology and Biotechnology	
Technology of Chemical Processes (Salonica)	
National Foundation of Research*	274
Institutes	
Biological Research	
Byzantine Period	
Greece in Modern Times	
Greek and Roman Period	
Organic Chemistry and Pharmaceutical Chemistry	
Theoretical Chemistry and Chemical Physics	
National Observatory of Athens	105
Institutes	
Astronomic	
Earthquake	
Ionospheric and Space Physics	
Meteorology	
National Center of Marine Research	138
Institutes	
Marine and Biological Resources	
Oceanographic	
Institute of Marine Biology, Crete	55

TABLE V. Continued from previous page

Research Center	Members
Pasteur Institute	114
Departments of	
Bacteriology	
Biotechnology/Biochemistry	
Immune System	
Virology	
Parasitology	
Vaccines	
Center of Biology A. Fleming	
Center of Research and Energy Sources	104
Lignite Center	
National Center of Social Research	107
Institutes	
City and Rural Sociology	
Economic and Social History	
Political Sociology	
Socio-economic	
Social Demography and Labor Relation	
Institute of Language Processing	45

*To this center also belongs the National Documentation Center that is having a large database and is connected with databanks in many countries.

The research centers play an important role in the system of science and technology. Certain weaknesses of these centers are given below:
a) the lack of systematic evaluation of research programmes and researchers;
b) the repetition of research activities in a number of research centers;
c) although they collaborate with each other and with institutes abroad, this collaboration is not satisfactory, particularly with one another;
d) no new research personnel is hired at frequent periods of time by the research centers.

Main goals for the improvement and the advancement of research centers are:

1. the collaboration and the co-ordination of research activities;
2. their closer connection with the production system;
3. efficient mechanisms for the evaluation of research programmes;
4. the increased collaboration with the international scientific community;
5. giving incentives for quality research;
6. clearly defining the specific goals of every research center.

Research in agriculture is supervised by the Ministry of Agriculture that is contributing the 14% of the funds and is co-ordinated by the National

Foundation of Agricultural Research which consists of a number of institutes distributed more or less uniformly around the country. Some of them are:

- the Institute of wine and grapes;
- the Institute of cotton;
- the Institute of wood;
- the Institute of wheat;
- the Institute of olive oil, forestry.

There is also a phytopathology research center. The Ministry of Education is supervising the institutes of the Academy of Athens, the Pedagogical Institute, the Institute of Technological Education and two institutes associated with the universities of Athens and Patras, the one in law research and the other one in informatics. Under a law that passed through the parliament recently, research institutes can be established by the universities or their departments and operate under their supervision with the rules of market economy.

Finally, the Ministry of National Defense is supporting three research centers but the funds allocated to them are limited. The Organization of Industrial Property Rights, the National Organization of Standardization, the Greek Productivity Center and the Organization of Small and Medium Enterprises, are under the supervision of the Ministry of Industry, Energy and Technology which provides the necessary support for research activities.

The General Secretariat of Research and Technology is the main office where science policy is formed. It supervises the research centers that have been already mentioned, the sectorial industrial research units, the National Organization of Industrial Property Rights. Also, the General Secretariat is responsible for the bilateral technological agreements with various countries, the representation of our country in international organizations, NATO, EU, OECD, United Nations, International Atomic Energy Agency etc. It is the main body for policy making with the European Union and co-ordinates all major activities related to R&D. Despite the limited personnel it is an efficient and flexible organization.

4. Conclusions

This was an overview of the Greek system for science and technology. The main goals to be achieved in the near future are:

1. the co-ordination of the various bodies that are involved in this field;
2. the setting up of priorities in certain areas of research;
3. the stimulation of interest in the private sector to become actively involved in research activities by providing the necessary funding;

4. the promotion of the collaboration of universities and research centers with the production system;
5. the establishment of science and technology offices in certain ministries i.e. Environment, Public Works, National Economy.

THE ROLE AND PLACE OF RESEARCH INSTITUTES IN ROMANIA

Restructuring Strategies between Tradition and Modernism

FLORIN TEODOR TANASESCU
Ministry of Research and Technology
Romania

The author presents the phases of the process of restructuring the R&D sector, the new mechanisms for financing research, competition as a basic principle, and participation in European research programmes. The directions in which Romanian research will flow in the years to come clearly start with the understanding of the new economic and social objectives. These are specifically reflected by the Resolution 1075/1995 of the Council of Europe concerning the co-operation of the UE countries with Central and Eastern European countries.

1. Romanian research before the Second World War

Before the Second World War, Romanian scientific research was mainly developed within universities that were concentrating the scientific community's life and were well known abroad, through their contacts with the great European centres of the time.

Romania brought meritorious contributions to the "world knowledge treasury", which had a prestigious place in the history of European science. The most representative fields of this performant research were mathematics, physics, chemistry, electricity, and medicine, in which some remarkable results were obtained, worthy to be mentioned.

In mathematics, a series of concepts and notions bear the name of their creators, being treasured by the world scientific community:

- Gh. Titeica discovers new categories of curves, surfaces and nets, that bear his name;
- Dimitrie Pompei introduces the notion of "aerial derivatie" and creates the "Pompei functions";

45

Ch. Proukakis and N. Katsaros (eds.),
The New Role of the Academies of Sciences in the Balkan Countries, 45–60.
© *1997 Kluwer Academic Publishers. Printed in the Netherlands.*

- David Emanoil contributes to the study of "Abelian integrals";
- Traian Lalescu develops the theory of integral equations, proving the existence of "periodical polygonal functions";
- Alexandru Pantazi defines, in the differential geometry, the "Pantazi quadruples" and develops, together with the Italian Terracini, the "Terracini-Pantazi nets";
- Gh. Mihoc brings important contributions to the development of "Marcov chains", mentioned in the specialized literature and introduces, together with Octav Onicescu, the notion of "complete bond chain".

In **the field of physics**, one can also mention some remarkable personalities and their achievements:

- Dragomir Hurmuzescu, trained in the Nobel Prize awarded physicist Gabriel Lippman's laboratories, determines the exact value of the ratio between the electrostatic and the electromagnetic unit and highlights the fact that X-rays give good electrical conductivity to the air.
- Stefan Procopiu discovers simultaneously with Niels Bohr, the theoretical magneton, which bears their name "the Bohr-Procopiu magneton".
- Nicolae Vasilescu Karpen demonstrates the existence of free electrons in liquids.
- Stefania Maracineanu is a forerunner in the discovery of artificial radioactivity, preceding by ten years the official certification of the phenomenon.
- Alexandru Proca establishes the equations of the mezonic vectorial field and, independently of the Japanese Yukava, postulates the existence of the meson, confirmed in 1937 by the American physicist Anderson, and establishes the "Price Equation".
- Theodore Ionescu establishes the theory of the cyclothermal frequency giromagnetic multiple effect, a phenomenon that was highlighted in 1962, following the observations made by the Canadian satellite Alouette.
- Horia Hulubei identifies, by spectroscopic methods, the chemical element numbered 87 in the Mendeleev Table, that was only inferred by Mendeleev under the name of "ekacesium".

Scientific research in the **field of chemistry** gave also some remarkable results at the European level:

- Constantin Istrate discovers the "franceines", nitrogen free, coloured substances, stable at light.
- Gh. Longinescu establishes "the relation of molecular associations" and gives a new formula for the Avogadro Law.

- Nicolae Teclu creates the "Teclu bulb" and is one of the forerunners of ozone industrial preparation.
- Lazar Edeleanu discovers a universal method that bears his name, to ensure extraction and selective refinery of crude oil aromatic hydrocarbides by using SO_2 (sulphur dioxide).
- Martin Bank develops the first patent in the world for obtaining acethylene, starting from methane gas.

Finally, I would like to refer to **the medical schools**, ending this presentation that was only aiming at highlighting a tradition of Romanian scientific research, that was laying the ground for other scientific disciplines that have been developed along the years.

- Daniel Danielopolu administers, for the first time in the world, small dozes of strophantine for the cardiovascular diseases, a method that remained in the medical practice under the name of "Danielopolu method".
- C.I. Parhon describes the hyperhydropexic syndrome, known under the name of "Parhon syndrome" and co-authors with M.A. Goldstein, the first endocrinology treaty in the world: "The internal secretions".
- Nicolae Paulescu discovers, in 1921, before Bonting and Mac Leod, the pancreine, the anti-diabetes pancreatic hormone, which is, in fact, insuline. The recognition of his priority in the discovery of insuline was only recently acknowledged.
- C. Levaditti and Stefan Nicolau discover, during their research work at the Pasteur Institute of Paris, the ultrafiltration capacity of viruses that allowed the further separation of viruses from bacteria.
  The cultivation, by Levaditti, in 1913, of the poliomyelitis filtrable virus, opens the way to the preparation, by the American Salk, in 1953, of the anti-poliomyelitis vaccine.
- Stefan Odobleja, in his work "La psychologie consonantiste" formulates the basic concepts of complex systems (biological, social), being considered the creator of psycho cybernetics and the Norbert Wiener's work forerunner.
- Victor Babes is one of the modern microbiology founders, with special contributions to the study of rabies, leprosy, diphtheria.

During this period of time, the place of scientific research is mainly in universities and the great scientific personalities acting here have founded valuable schools, have approached modern topics and laboratories, proving a real capacity in approaching the new challenges.
One can also talk about applied research in the industrial units that are now created in Romania, in the field of oil extraction and refinery, electricity applications in industry, traction, lighting, aviation, metallurgy.

2. Romanian research between 1940s-1990s

This period is dominated by a rapid development of Romanian industry in various economic fields.

The growth rate and the transfer of know-how to top industries, such as: agricultural machines, aviation, petrochemistry, semiconductors, electronic devices, and electrotechnique, generated a growing interest for research.

The strong development of technological research, approaching both basic research and applied research, led to a Romanian "research market", in the sense that research became an extremely tempting option for young graduates.

By the end of 1989, some 200.000 persons were working in technological research and factory research departments, out of which some 70.000 persons were graduates of higher education institutions.

This fact proves that numerous research teams were increasingly consolidating, they were developing many important research projects on their own conception and design and were setting professional priorities that gave an important support to the development of technical skills.

The drilling industry (deep drilling equipment or off-shore platforms), the subway of Bucharest, achieved entirely by Romanian conception and industry, airships (YAR-99, YAR-93), catalysts and synthesis processes in petrochemical industry, automation and informatics, new animal breeds and plant sorts, the successful achievement of a technology for the obtention of heavy water, all by Romanian conception, as well as equipment for a nuclear power plant, the Danube hydropower systems of Portile de Fier or power systems of other inner basins, are only a few domains where these successful results were backed-up by a mature Romanian research.

Now, looking back from a distance, one can better assess the positive and the negative influences that, to a certain extent, were reflected in the history of Romanian research.

2.1. ELEMENTS OF POSITIVE INFLUENCE ON ROMANIAN RESEARCH

- An alert industrial development brought with it an important demand for research, in the sense of an effective and direct involvement of research in the process of new industrial technologies creation.
- Important financial resources could be dedicated to the development and endowment of scientific laboratories.
- Although it was placed in a different economic system than Western European countries (namely the centralized economy and the

COMECOM market), Romanian research kept its contacts with the great European Scientific Centres and their scientific trends and applications.

For instance, although Romania was a part of the COMECOM system, the Romanian standardization in electronics and electrotechnique was achieved according to the standards of the International Electrotechnical Commission, and not according to the COMECOM norms.
This fact explains also the relatively easy harmonization of Romanian quality standards with the European ones.

- The existence of research funds, either from industry or from the state budget, is able to generate an increase of the research personnel number.
- A closer connection between technological research and university research, even if, in some cases, is difficult to achieve, was the alternative for a modern research, which is actively promoted nowadays also by European countries and represents a guarantee for the improvement of the industrial product performance level.
- Given the large industrial demand for research, many research results had great possibilities to be applied.

2.2. ELEMENTS OF NEGATIVE INFLUENCE ON ROMANIAN RESEARCH

- Low interest for university research and for the development of long-term research projects by the university chairs.

This led to a modest endowment of university research laboratories, which now raises great problems for the actual research management.
In the years to come, important efforts have to be made for the modernization of the university research laboratories. If this endeavour is not properly done, it may lead to a limited participation in the most important European research programmes.

- The abolishment of those Romanian Academy institutes which are active in technical fields and which their merger with other research institutions had a negative impact on the number and perspectives of long-term research works.
- Although the concept of microproduction development in research institutes had positive structural elements in the sense that the author-researcher could check his solution before offering it to an industrial beneficiary, the exaggeration of this concept by oversizing the microproduction, led to an increase of technical and auxiliary personnel in these institutes and to a blockage of valuable human resources in this sector.
- No participation in European research programmes and restricted international mobility of researchers, with poor chances to travel to

COMECOM member countries and even poorer chances to travel to industrially developed countries.

- Difficult access of young graduates to research institutes, which became even more difficult in the late 1980s, which explains the relatively old age of qualified researchers in many institutions.
- The lack of adequate incentives for the promotion of researchers, as well as for the transfer of research results to industry.
- Many obstacles faced by scientific research, caused by the reduction of components, materials and know-how imports from the developed countries, imposed increased efforts for scientific research to provide local solutions for their replacement.

3. Some R&D statistical indicators for the period 1990-1996

The evolution of the number of R&D units and their personnel during 1990-1995 is shown in Figure 1.

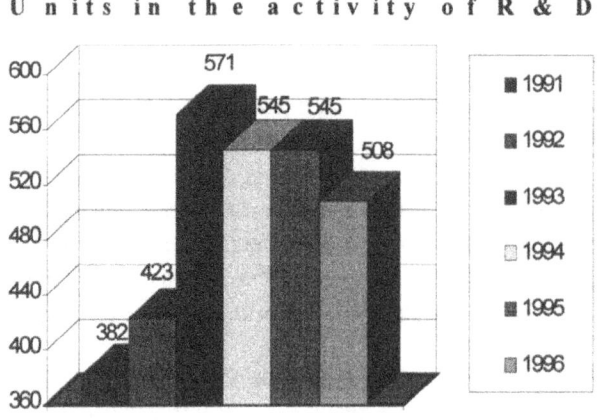

Units in the activity of R & D

F i g . 1

Typical for this period was an increasing number of R&D units generated by the splitting of some big institutes into smaller and more flexible spin-offs.
As far as research personnel is concerned, as a result of the restructuring and departure from R&D institutes, the personnel was 118.000 in 1995, out of which 48.500 had higher education degrees (Figure 2).

Employees in the activity of R & D

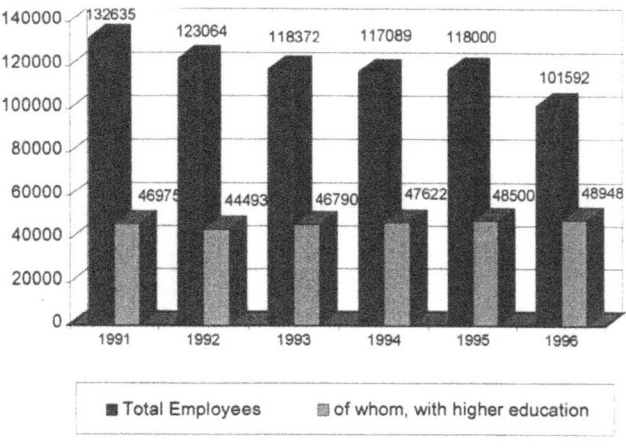

F i g . 2

Starting with 1993, one can observe a slight increase in the number of highly-educated researchers.

Figure 3 shows the evolution of researchers by age groups in Romanian R&D units. One can notice the high percentage of the 40-60 age group, which might be improved in 5-10 years time, provided that adequate incentives for young researchers will be adopted.

Researchers by groups of age

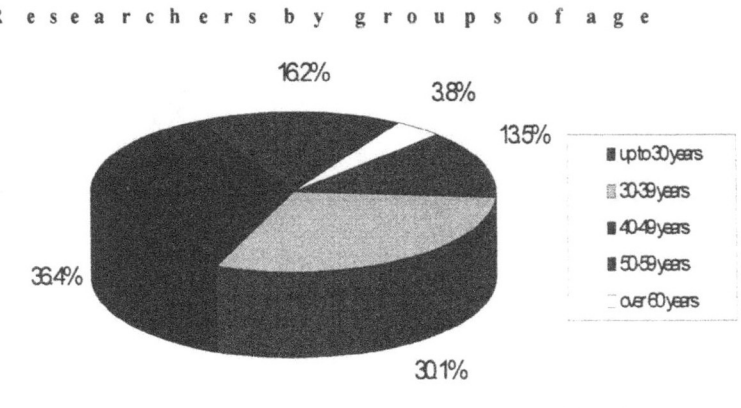

F i g . 3

Figure 4 shows the total number of R&D personnel per 100 full time employed persons. Although important losses have taken place due to brain-

52

drain, the total number of R&D personnel compared to the total number of employed persons in other countries is satisfactory, taking into account the situation in some European Union countries, as Portugal, Greece, Spain.

R & D   P e r s o n n e l (%   o f L a b o u r   F o r c e)

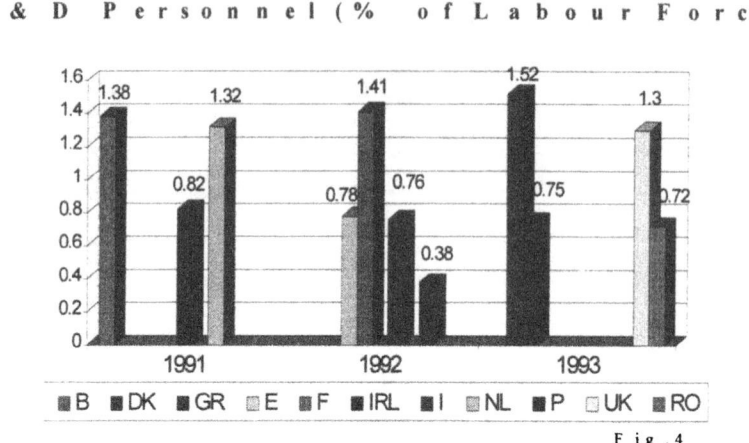

Figure 5 presents the professional R&D qualification, showing a satisfactory number of PhD's and PhD students.

N u m b e r o f D o c t o r s a n d D o c t o r a n d s in R & D u n i t s b y a g e

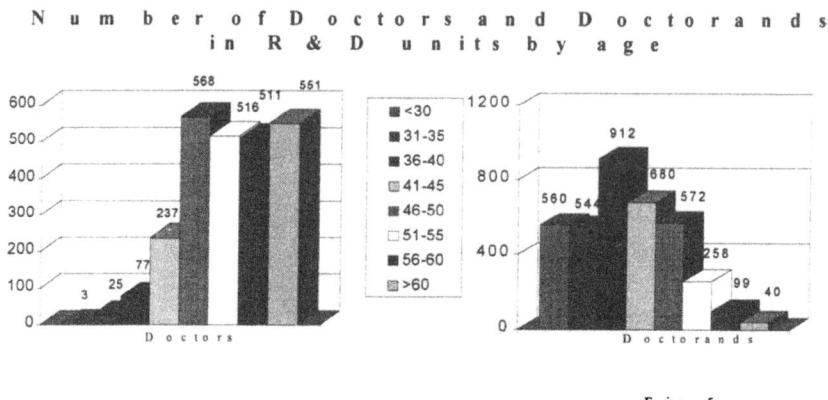

F i g . 5

Figure 6 shows the total R&D expenditure during 1990-1995, both from the state budget (including the 1% special fund for research, until 1994 when it was abolished) and from the economic agents (continuously increasing).

R & D Expenses in G D P (%)

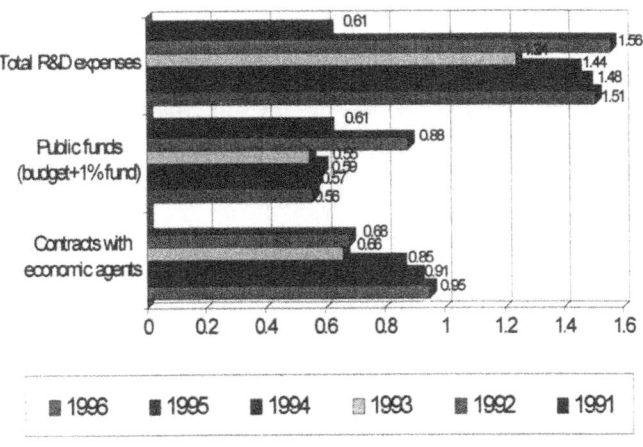

F ig .6

The introduction of a co-operative research principle and of co-financing had a positive influence by increasing the number of projects managed by joint research teams.

The percentage of R&D expenditure from the GDP is shown in Figure 6. One can also remark the increasing percentage of direct contracts with economic agents. Figure 6 shows also the total R&D expenditure by different funding sectors.

Figure 7 presents the evolution of funds for the projects coming from university research teams and industrial research centres, especially SMEs, which the Ministry of Research and Technology is providing as a "catalyst" in the scientific community's life.

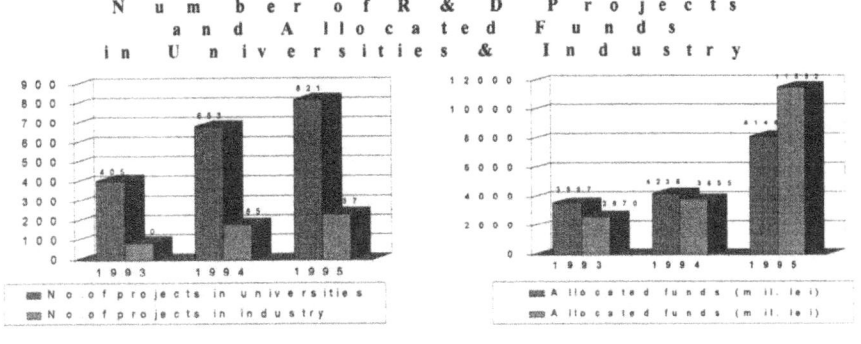

F ig . 7

54

The decentralization of scientific research, in some regional centres, led to the identification of effective solutions to the specific local problems and stimulated the innovation activity and the number of inventions.

Accordingly, the number of research projects in this field funded by the Ministry of Research and Technology grew continuously year after year, and the **Romanian participation** in the international exhibitions of inventions was rewarded with a large number of medals and special prizes, as shown in Figure 8.

Support for innovations

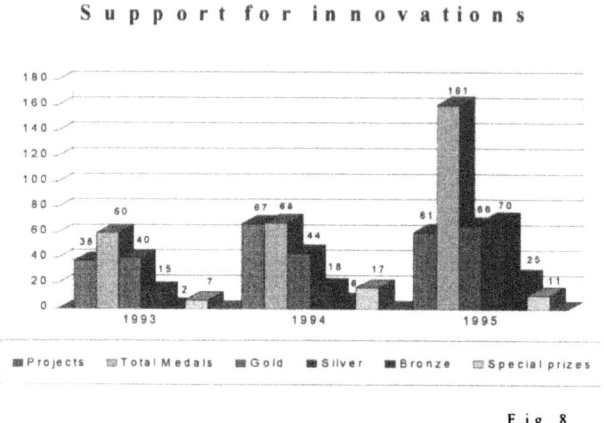

Fig.8

Figure 9 presents an evolution of the number of patents, which **strongly decreased before 1992**, slightly increased after 1992, but without reaching the level of 1988.

Patent Applications

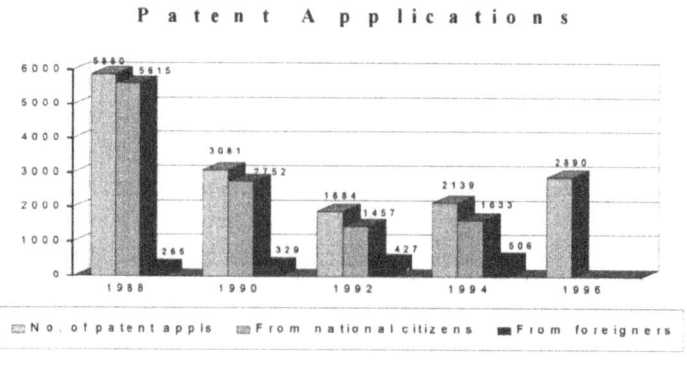

Fig.9

Figure 9 estimates the repartition in 1996, by different types of institutional framework, of R&D units.

4. Research restructuring strategies after 1990

After a brief presentation of some structural elements, concepts and statistical data concerning Romanian research of the last 6 years, let me say a few words about the restructuring strategies that needed to be adopted in order to shift from a centralized economy to a market economy, strategies aiming at adjusting the existing structures, domains and mechanism, to the European ones (Figure 10).

4.1. ROMANIAN R&D CHANGES DURING 1990-1994

The main changes that took place during this period were:
- abolishment of co-ordination bodies of the former system, definition of the role played by the new institutions called to provide a strategy of research in Romania, the autonomy of R&D units;
- autonomy of the Romanian Academy, the Academy of Medical Sciences and the Academy of Agricultural and Forestry Sciences;

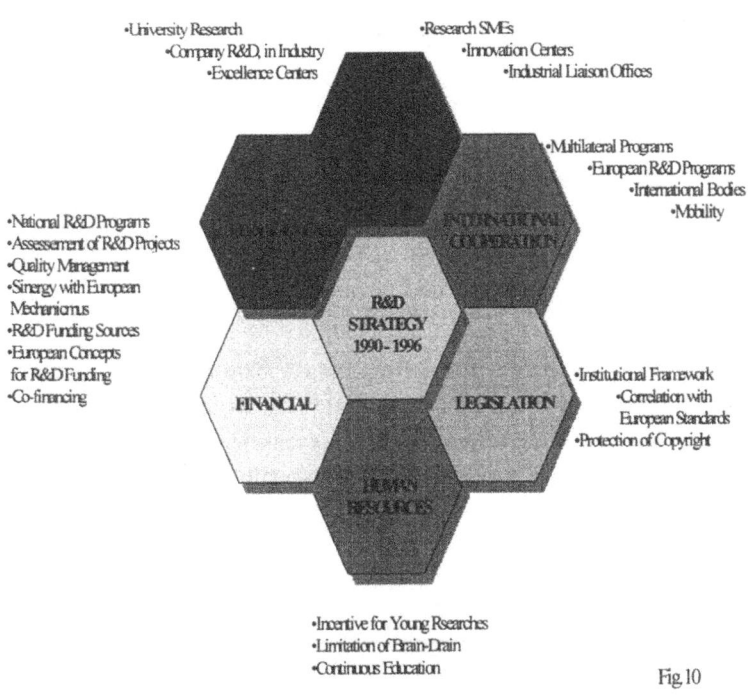

R&D STRATEGY 1990-1996

Fig.10

- autonomy of R&D units, both in terms of organization and functioning;

- the creation of a R&D special fund (1990-1993), based on a 1% tax on the economic agents' turnover.
  I would like to insist upon this matter because it had a great importance upon the very existence of Romanian research. The Government realized that in this period of economic crisis, generated by the transition process - with a painful reduction by 50% of the GDP - the demands for research will not be significant. Therefore, the Government established this 1% tax fund, which has functioned for a three year-period and offers to scientific research, an opportunity for restructuring, priority setting, according to industrial restructuring, and for stopping the undesirable "brain drain" of the period 1990-1993, (when scientific research lost 35-40% of its personnel, which shifted to private sector or abroad).

- the creation of the **Advisory Board for Applied Research and Technological Development (in 1991),** a co-ordination body of the Romanian scientific community, industry, education and Romanian Academy, whose role is :
  i) to draw up the national R&D programmes,
  ii)to grant, on a competition basis, the funds for R&D projects;

- **the establishment, in 1992, of the Ministry of Research and Technology,** whose role was defined as follows :
  to draw up the scientific research strategy and to find the best ways and means for its promotion;
  to be an equally distant arbitrator for all the research sectors (in higher education, in R&D institutes, in the Romanian Academy, in industry), stimulating them in meeting the demands of the national R&D programmes;
  to be involved in R&D restructuring and reform;

- the creation of **the Interministerial Council for Science and Technology (in 1994),** a governmental body which harmonizes the R&D activity, developed by the different research units active in different fields, assists the government in adopting some regulations and provides a coherent national research and technology policy.

As general strategic objectives for the science and technology policy after 1992, the following targets have been defined:

- encouraging sustainable development of the country on the basis of innovation in economy and society.

- accelerating the process of Romanian research integration in the E.U. research.
- making research capable to meet the challenges of the Information Society.

The period 1993-1995 brought the restructuring of the main components of the R&D system at the institutional, managemental, financial, human resources, legislation and international cooperation levels. In the Figure below, the main actions related to these changes are presented.

On the basis of these general strategic objectives, the national R&D strategy for the period 1996-2010 was set up.

5. The national R&D strategy for the period 1996-2010

For the period 1996-2010, the aims of national R&D strategy may be defined for shorter or longer terms, but they can be affected by a series of tendencies which appear today in the world, and for which solutions are still looked for: industry, markets and research globalization, information society, a.s.o.

Short term strategies (1996-2000):

- reinforcement of partnership links between the research & development sector and economy.
- increasing the competitiveness of R&D human resources by setting and applying stimulating policies for professional promotion.

Medium-term strategies (2001-2010):

- increasing the R&D funding level by a more important industrial contribution;
- setting up a competitive local market for R&D;
- creation of enterprises of a scientific, production and commercial profile, of SME - type;
- rapid application of R&D results in products and technologies;
- reaching the excellence level according to international standards;
- the extension of scientific and technological partnership at an international level, and the affiliation to the best international research laboratories;
- the effective integration with the global trends;
- the consolidation of society, based on innovation.
- the Romanian research priorities for the next period of time, will be those mentioned in Figure 11.

Fig.11

Romanian Research Priorities
☞ *The life sciences (biotechnology, health, food and agricultural research)*
☞ *Informational technologies and telecommunications*
☞ *Production, process and material technology*
☞ *Supporting SMEs and Promotion of Technology Transfer*
☞ *Environmental research and technology*
☞ *Social and cultural phenomena with a European dimension*
☞ *Relating Romanian R&D to European Research and Centers, accomplishing synergy between national R&D Programmes and the fifth Framework R&D Programme of the European Union, with other scientific Euro-Atlantic Programmes*
☞ *R&D programmes for the protection and development of the Danube Delta and The Black Sea ecosystem*

6. Some remarks, instead of conclusions

The restructuring of scientific research in Central and Eastern European countries was modelled according to today's organization and functioning of European research and not to tomorrow's one.

Taking into account that even developed European countries, facing these big development problems, search for new organization forms of the R&D activity in the transition countries, it is possible that this search by us be extended much longer than expected, with the respective implications.

- Authorized voices try to explain the lower competitiveness of European countries as against the USA and "the Triad" due to the weaker links between the research units and industry, the slow transfer of some solutions, the absence of the harmonization between the funding level of national research programmes and European research, compared to Japan and US.
- The globalization of industry and markets will naturally bring a globalization of research.

Unfortunately, we do not know yet which will be the most suitable structure, the most adequate mechanism.

A series of questions rest without answers today, but they must be found as urgently as possible!

- ◊ What mechanism must be adopted to approach research in university, industry, society, as it is expected?
- ◊ How would one best connect the research centres from different countries to European programmes?
- ◊ How can we increase the risk capital in the high technology sector?

◊ What less restrictive rules will have to be taken, for the research activity and technological transfer to offer an incentive to society?
   And what financial stimulative policies will have to be promoted: income-tax reductions, guarantees and credit schemes, stimulative mechanisms for technological transfer, partial project funding?
◊ How could one attract the small and medium enterprises to the creation process, to the innovation-based development?
◊ Promoting a real culture of innovation, how could one best "articulate", in this process, research - industry - education - society?
◊ How could the trans-European capital be attracted to innovation funding?
◊ How flexible must the European programmes be to face the challenges?

In a recent interview given by the European Union Commissioner, Mrs. Edith Cresson at the Bonn University (16 Oct. 1996), referring to the fifth Framework Programme of the European Union (1999-2002) and insisting on the problems which will impose new approaches, Mrs. Cresson mentioned a few specific features of the stage that research must keep count of:

* the increase of the social demand for research, and the interest of society not only for "outrunning the unknown borders", for "the human spirit honor", but also for finding concrete and rapid answers to the societal problems;
* the profound transformation of economy and the nature of work, at the same time with the access to the "immaterial economy", the real richness laying in adequate exploitation of information and knowledge.
* the necessity that future research programmes be more targeted, more project-oriented.

The necessary shift should be from a research which was based until now on the technical performance, to a much better oriented research of the socio-economic necessities, human aspirations, in health, environment, life quality matters, improvement of knowledge in "key" fields for the future, the setting up of the favourable climate for research; stronger support for the actions which lead to a better connection between economy-technology-culture-work place; SME' s access to research.

These are a few message-ideas, which anticipate substantial changes of research structure and which one must take into account.
A few hundred years ago, the philosopher Francis Bacon gave a visionary picture of science organization in his book "New Atlantis", in which "the House of Solomon" was based on research organization, experiment planning, and team research superiority compared to individual research. Many of his ideas are found today in our way of thinking, associating and working.

The fifth Framework Programme of the European Union was recently discussed at the meeting of the Research Ministers of the EU member states, held on October 7th, 1996. It emphasized a human message: "Inventing Tomorrow - Europe's Research at the Service of Its People". It's a challenge addressed to the scientific community in order to find a way to reinforce the future and the society it is serving. Any kind of contribution, wherever it comes, will be welcomed.

I hope we have already begun to do it, now, together, by carrying on this interesting seminar.

THE ROLE OF THE NATIONAL ACADEMY OF SCIENCES IN THE SCIENCE AND TECHNOLOGY SYSTEM OF ARMENIA

YURI L. SARKISSYAN
State Engineering University of Armenia
105, Terian Str., 375009, Yerevan, Republic of Armenia

1. Introduction

In [1] some guidelines for restructuring the science policy framework in Armenia are presented. With the disintegration of the centralized Soviet research system, the main policy concern for the country has become the survival problem of the existing powerful science base and its reorientation to the new national goals. On the other hand, it is obvious that without a vigorous science and technology system the country cannot enter the mainstream of the global economy and build a competitive technology driven industry. The point is how to combine the short term preservation measures with a long term strategy for building a S&T system adaptable to the emerging market economy.

The paper is concerned with the transformations of science policy and research management institutions in Armenia, focusing on the new role of the National Academy of Sciences in the S&T system.

2. The present state of the science system

The network of R&D performing institutions structurally consists of 3 distinct organizational entities:

a) more than 30 scientific institutes of the National Academy of Sciences;
b) the laboratories and other research units of 14 universities and other higher educational institutions;
c) the Branch R&D institutions of different Ministries oriented to industrial research.

The coordination between the academic, university and industrial R&D sectors has been achieved in the past by the party and state coordinating bodies at the central, regional and institute levels. After the dissolution of the

61

Ch. Proukakis and N. Katsaros (eds.),
The New Role of the Academies of Sciences in the Balkan Countries, 61–68.
© *1997 Kluwer Academic Publishers. Printed in the Netherlands.*

former "political" coordination system, some new and efficient coordination structures were needed to integrate the efforts and potentialities of three disconnected science communities in achieving the new national goals. Some important organizational and policy changes have been made in S&T management and funding since 1990, when Armenia received independence and started major economic reformations. The initial step was the establishment of the Ministry of Education and Science and the Department of Science and Graduate Education as a division of it, with the functions of general coordination of funding, research management and realization of science policy throughout the country. In addition, a Special Foundation of Science and Innovation was organized for the alternative funding of the new projects, which was soon dissolved for the lack of necessary resources.

In 1992, a new and unified research funding system was introduced which was project based and rested on the peer review process implemented under the joint responsibility of the Academy, the Ministry of Education and Science and the Ministry of Economy. The science budget share in the 1996 total national budget represented 1.35% amounting to $ 4.5 million which was distributed through the peer review process between the three R&D sectors by the following properties: academic system 50 %, university research 23 %, branch institutes 27 %.

The new supporting scheme had, of course, some positive and stimulating influence on the creative initiative and freedom of the researchers and research groups, particularly on those that did not belong to traditional R&D institutes or labs. But the new funding scheme lacked appropriate project evaluation mechanisms capable to differentiate between the basic research projects and the projects from the domain of applied and industrial research which require different evaluation criteria and methodology.

Moreover, the new research grant system excluded the traditional institutional and program funding schemes and as a result broke up many distinguished research institutes and labs into a mosaic of small units lacking coordination. The academic research institutes have been most seriously affected by these changes.

To fill the vacuum in science policy formulation and policy making structures, the Council of Science and Advanced Technologies was established in 1996, chaired by the Prime Minister of Armenia, with the powers to prepare the National R&D strategy, initiate national priority programs, assist the government in defining the level and the structure of the annual science budget and supervise its implementation which had a wide representation from the main research institutions and sectional science communities, including the Academy and the Universities. The Council has not formed yet a reliable organizational framework for the program development and project evaluation and specialized sectional committees to prepare strategic recommendations and evaluation reports for the different branches of science and technology.

3. Guidelines of the science policy development

The main S&T policy issues are the following:

- *The development of the national S&T strategy*

Here the main concern is to declare, at the state level, science and technology as one of the key national priorities, taking into account the strong science base, traditions, and potential impact on the economic development, reactivation of industry, quality of life, including environment and health protection. The focus of the strategy is the identification of the priorities in the field of R&D and special R&D programs addressed to these priorities. The criteria for the identification could be the strategic directions of the economic development, priority technologies to be revitalized or developed, competitiveness on the international level. The S&T national strategy should be developed by the joint effort and consensus of the government, scientific institutions and S&T communities and be initiated and coordinated by the Academy.

- *Formulation of the legal framework for the S&T policy*

Armenia does not possess yet any law on S&T which could fix the distribution of powers and responsibilities between the governmental structures, legislative, organizational and financial mechanisms of the state influence on science and technology activities.

- *Adaptation of policy making institutions and research funding system to the new goals of the S&T system*

A research management structure (Department or Ministry) responsible also for technology development is needed at the government level as the executive body of the Council of Science and Technology to coordinate R&D activities of the academic, university and branch institutes and to implement the strategies and decisions of the Council. But the key issue here remains the establishment of an appropriate financing system, the conditions under which research is funded, and the administration of the funding process. It is necessary to establish all three main channels of funding: institutional funding, program-based funding and project-based funding and harmonize their shares in the state science budget.

- *Evaluation of Scientists and Institutes*

The drastic reduction of the science budget, which is the direct result of the transition problems, requires to select the best institutes and scientists and to

64

concentrate the support on them. This selective funding approach is crucial especially for defining the priorities of institutional funding since the science system of Armenia is obviously oversized and should match with the new economic realities. In this situation, reliable evaluation procedures, including peer review mechanisms, are of cardinal importance to preserve the most competent research teams and projects.

- *Restructuring of R&D performing institutions*
The development of rational reduction and reorganization plans within the different R&D institutional structures, in response to the changes of the economic environment, is also one of the key policy issues. As in every ex-Soviet Republic, R&D is structurally disconnected from education and industry and now one of the guidelines for restructuring the R&D institutes is the development of cooperation and integration mechanisms with the universities and technology oriented enterprises.

4. Reorganization of the Academy of Sciences

The Armenian Academy of Sciences was one of the most advanced and internationally recognized science systems in the former Soviet Union and used to have very high prestige and institutional status in Armenia. It was a close looped powerful organization, a kind of "scientific" Republic in Armenia which administered independently its own network of institutes and funds and had a significant influence on scientific policy setting of the country. At the same time, the Armenian Academy, organizationally and financially, was closely linked to the Central Soviet Academic complex, being a component of it.

Due to the unique position of this institution in Armenian science, the status of the Academy in the new S&T system has become naturally one of the major science policy issues most disputed recently at the government level. In 1991, a state Ad Hoc Commission was formed by the decree of the President of Armenia to prepare recommendations on the reform of the Academy.

The focus of the discussions was the new status of the Academy. According to the Soviet model, the Academy used to combine two conceptually different missions:

a) awarding honorary degrees to Academy members (members and corresponding members) through the elections based on the evaluation of the scientific results and merits of the candidates
b) management and administration of the Academy research institutes.

Historically, this dual mission brought to an eclectic organizational structure of the Academy in which two principally different entities were put together: the honorific part (members of the Academy) and the operative part (academic research institutes and labs). Moreover, the science administration functions were implemented by the governing bodies - the Presidium, the Central and Divisional General Assemblies, composed exceptionally of the Academy members, while the research institutes did not have any direct representation in these bodies or any participation in elaboration of the administrative decisions. In this scheme, the Academy members elected only for their science merits but not for the administrative skills which concentrated in their hands the main powers of controlling the funds and deciding the research priorities. This model had, of course, a definite restrictive influence on the initiative and progress of the researchers and research teams.

The focal question which had to be answered was whether the Academy will be reorganized as an Honorific Society of academicians without science administration functions or it will continue to manage its network of research institutes as before. The principal recommendation of the Commission was to preserve the organizational unity of the Academy research system and maintain its leadership in the supervision of basic research at the national level. In addition, some major organizational changes were suggested to separate the honorific and operative components of the Academy, to improve the management and coordination of the research units, and to ensure the adequate representation of the research institutes in the main governing bodies of the Academy.

The Commission proposed also to dissolve the two-level membership in the Academy (members and corresponding members) which had a negative effect on the academic environment and quality of the election process, and establish a single rank of the Academy members.

Based on these recommendations, proposals of the Academy and discussions in different science communities, the President of Armenia issued in March 1993 a special Decree on the reorganization of the Academy. It defined the status of the National Academy of Sciences of Armenia as a self-governing state scientific institution, uniting the members of the Academy, its foreign members and the research staff of the Academy system.

The Presidential Decree defined also the new mission of the Academy as the official adviser to the government on S&T policy and the general coordinator of basic research in the country, its powers and responsibilities in the development and implementation of the priority national R&D programs and strategies, project evaluation procedures and peer review mechanisms, R&D legislation. Shortly after, the new Academy Statute (Charter) was ratified by the government which was based on the principles fixed in the Presidential Decree.

The Charter provides for the new structure and management system of the Academy with the following characteristic features:

- The Academy research system has a tripod structure and is composed of three Divisions: Division of Natural Sciences, Division of Informatics, Physics, Mathematics and Technical Sciences and Division of Humanity and Social Sciences.

- The Divisions are chaired by the Sectional Academic Secretaries who combine the duties of the Vice Presidents and are responsible for the development and supervision of the corresponding branches of science in the Academy.

- 8 discipline/problem-oriented Councils have been set up which cover all the specific research areas of basic sciences. These councils, comprising the leading experts and scientists from each discipline area, are the supporting structures for the Academy to implement its advisory and coordinating functions assigned by the Charter.

- The restructured General Assemblies and their Divisions include mandatory (ex officio) members - the academicians, directors of research institutes and elected members - representatives of different research institutes.

- The institute of the corresponding members is dissolved and a one-level membership of the Academy is established.
  There are some strong internal (institutional) and external (environmental) factors which restrict further reformations of the Academy and efficient implementation of its mission and responsibilities.

- Since the channel of institutional funding is closed, the Academy does not have any direct participation in the distribution or control of its research funds, granted on a project basis to the research teams. This does not permit the creation of any long-term development strategy for the research institutes or the launching of special purpose R&D programs and new research activities.

- The newly established Council of Science and Advanced Technologies has not created yet the necessary interaction mechanisms with the Academy to ensure the efficient implementation of its advisory and coordinating functions.

- The Academy research system is isolated from education and industry. It is appropriate to try to link the Academy institutes closely to the universities by the associative organizational structures or formal agreements of cooperation, including the joint graduate degree programs with the application of the Academy Science base.

- The new Statute does not contain any new and coherent with the environmental changes solutions for the improvement of the management and administration of the Academy research system. The honorific and operative parts of the Academy are still coupled in the same governing structures and need immediate decoupling to give operational freedom to the institutes. The idea of creating National Research Centers, bringing together the research institutes from the same area of science, could be one of the possible directions of reorganization.

5. Conclusion and basic recommendations

Armenia is currently engaged in a vast transformation in which science and technology are one of its strongest assets. The country needs to restructure its S&T system in a way that will contribute best to the economic and social development. Although the general strategic requirements for restructuring S&T are equally admitted by the government and science communities, the overall transitional environment is not favourable to the changes. After the reorganization chartered by its new Statute, the Academy of Science maintained its leading role in Armenian science, as well as the organizational unity of its research system and legally confirmed its status of the state scientific adviser and coordinator of basic research. Nevertheless, there are some problems of an institutional and environmental nature which limit the implementation of the new mission of the Academy. First, it is obvious that the Academy can not be established as a self-governing institution and maintain its influence on S&T without appropriate institutional funding and its own funding mechanisms. Second, to build any development strategy for the Academy, its research budget share should be defined within the state annual science budget. Third, it is impossible to guide S&T and its institutions, particularly the Academy institutes, from the old system to the new without appropriate legal and organizational framework for the reformations.

The following principal recommendations may be drawn from the preceding analysis:

- In the situation of a strong science budget reduction, when downscaling of the Academy research system is inevitable, the development of a rational reduction and reorganization plan within the Academy is necessary to avoid unplanned destructive changes. The inclination of the Academy to distribute the available scarce resources equally does not help to retain the best science potential. The first task is to define investment priorities by

68

evaluating the accumulated scientific capacity in different research fields and then use the reorganization instrument for promoting the vitally important elements of the system and eliminating its ineffective parts.

- To make the necessary choices in a judicious way and to start the selective funding process, reliable evaluation mechanisms should be developed for scientists, projects and institutes. The introduction of peer review mechanisms with international participation has proved to be the best approach so far for basic research and is equally applicable to science personnel, projects or institutes.

- If the Academy wishes to remain effective in a new and unstable environment and fulfill its statutary mission, the institution should be sensitive and adaptable to the changes of the environment. A systematic environmental scanning is necessary to identify the main external and internal (institutional) factors (or variables) which may have significant influence on the future of the Academy. This will allow us to find out how to control best the environmental effects. The future Strategic Plan of the Academy should combine the understanding of environmental trends and opportunities with the analysis of its strengths, weaknesses and possible options for reorganization to produce strategic aims and objectives to be achieved.

References

1. Sarkissyan, Yu. L.(1995) Guidelines for Science Policy and Research Management in Armenia, in G. Parissakis and N. Katsaros (eds.), Science Policy and Research Management in the Balkan Countries, Kluwer Academic Publishers, Dordrecht, pp.165-171.

RESEARCH AND THE EVALUATION OF ITS RESULTS IN INDUSTRY

G. PARISSAKIS
National Technical University of Athens
9 Heroon Polytechniou str., 157 73 Athens, Greece

In the present strongly competitive technological environment, the improvement of products or the development of new ones have proven to be essential elements for the survival of an industrial activity.

Therefore, a question is immediately raised. How such an activity and more specifically a technological improvement or development can be set up, in order to achieve its goals.

Before though a decision is made, some facts and data must be necessarily taken into consideration, in order to help and sustain the actions to the right direction.

One of the most difficult and crucial steps in this process is to evaluate the most realistic perspective on how a new product or an improvement of an existing one can influence positively the prosperity of a company. And speaking of the prospects of a company, let me declare without any hesitation that in the present rapidly developing technological activity, no product, except maybe only a few, can compete successfully for a long time period with similar products of a more recent generation.

The lifetime of a new product or a technological improvement of a process must not exceed a period of five years. During this time period some corrective interventions may be proven necessary in order to remain in the forefront of the developments.

The team which will be involved with the creation of a new product, or with the improvement of a manufacturing process, must consist of scientists of not only high technological level and specialization in a specific industrial activity, but also towards other activities as well, such as marketing, product development etc.

The effort for improvements must be continuous and include even products which are considered as traditional ones, or others based upon a standard technological procedure.

Ch. Proukakis and N. Katsaros (eds.),
The New Role of the Academies of Sciences in the Balkan Countries, 69–73.
© *1997 Kluwer Academic Publishers. Printed in the Netherlands.*

At this point, I will refer to a characteristic example of an industrial product, that of the grinding balls, which are used in very many grinding and size diminution processes.

A characteristic phenomenon of these balls is their wear, and accordingly their shape deformation during the grinding process, resulting in their failure to accomplish their use.

Until some years ago the traditional way for the manufacturing of these balls was based upon the use of an appropriate alloy of specific characteristics such as hardness, brittleness, etc.

Nowadays, some technologically advanced steel manufacturing industries, produce grinding balls such that, their wear is about five times less of those of the traditional type. This new grinding media was the result of a successful research activity that led to the development of multi-layer balls, meaningly balls consisting of layers of alloys with different characteristics, thus resulting in their reduced wear.

It is pointless of course to stress that the older technology grinding media are almost set aside by those of the newer technology.

And when a thorough study concerning the development of a new product or the improvement of its characteristics has been completed , then comes the choice of the most appropriate procedure in order to achieve the desired goal.

There are many different procedures in order to obtain the necessary knowledge to improve an industrial process or create a new product. Of course all these have disadvantages and advantages, some of them may look very attractive or very expensive at a first glance.

Thus, the choice of the most appropriate and convenient procedure is the second very important step in the decision making process.

In the following table the most important of the existing procedures are listed (not in order of importance).

1.	Purchase of "Know-how".
2.	Purchase of a patent.
3.	Technology transfer with or without participation of the owner of the technology
4.	Contract for a research project with a specialized Public Organization or an Academic Research Institution
5.	Contract for a research project with an Independent Research Organization
6.	Contract for a research project with a Specialized Laboratory belonging to a group of Companies which are active in the same field
7.	Contract for a research project based on the ability of an R & D Laboratory owned by the company itself.

Let me now discuss in detail those possibilities.

1. Purchase of know-how - Technology transfer

A main prerequisite for the evaluation of know-how is to what extent the knowledge deriving from it or through transfer of technology can be incorporated in an already existing process.

Thus the incorporation to an existent manufacturing process or the production of a new product, without the assurance of the ability of further development based on the capacity of the company's staff itself is an imprudent action.

2. Patent purchase

Purchase of a patent in order to be used in an existing procedure is a very difficult decision with a lot of unforeseen aspects.

The age, for example, of the different novelties described in the invention as well as the confirmation of all the data mentioned in it and to what extent some economic aspects can be considered as realistic, are some of the important criteria for its purchase.

It is well known that the data mentioned in a patent cover a wide range of arithmetic values, while very few of them are functionally correct and realistic.

So, in many cases, it is very difficult to estimate the correct value of a patent and accordingly, to my own experience, this is a fairly complicated problem with an unforeseeable final result.

3. Technology transfer with or without participation of the owner of the technology

What has been mentioned in the case of "Purchase of know-how" is valid in a great degree for the technology transfer as well. A tight collaboration between the owner of the technology to be transferred and the receiving enterprise is an essential element in the process of implementation of the new technology, at least for the first steps of collaboration.

4. Contract for improvement of an existing process or production of a new product with a specialized Public Organization or an Academic Research Institution

Provided that discretion can be assured, this way for buying new technology or a manufacturing procedure for a new product seems to be in many cases very attractive and a very promising one.

The high technological level of the Public Research Organizations or the Academic Research Institutions is a very essential element towards the success of this kind of contracts.

The exchange of information and ideas between the scientific teams of the two parties on an almost daily basis is a priori a very important and positive aspect of such research contracts.

5. Contract for the establishment of a new process or for the set-up of a new product with an Independent Private Research Organization or Laboratory

This way, very common in many European countries as well as in the USA, generally gives very positive results. Dissemination, in this case, of the results is almost impossible because these institutions are greatly concerned about their credibility , given that , being active in a free market environment , they risk the chance , in case of not keeping the necessary secrecy , to start losing their clientèle.

In addition to this, the fact that the scientific, technical and managing staff of those private research centers is bound by a contract to keep secret the targets and results of the projects, appears as a very important and at the same time very reassuring element.

Thus, the product of the research contract belongs exclusively to the one who has financed the project.

Generally speaking, these research contracts are more expensive than other previously mentioned, but are proven very reliable given the professional state of these institutes.

6. Contract for a project with a Specialized Laboratory belonging to a group of enterprises which are active in the same technological field

This is a procedure followed sometimes, by smaller companies which can not afford to incorporate in their activities a research team or a R&D laboratory.

In such a case, of course, discretion is not possible and so these laboratories are used mainly to solve production problems or to control a standard production. Their everyday occupation with subjects concerning the branch in which they belong, gives them the ability to consult enterprises of this branch and inform them about trends and perspectives of future activities.

7. Research projects based on Research and Development Departments of the Company

This case fulfills the demand for discretion, reliable solutions which are based on their own experience, the set up of very defined targets and a fairly quick solution of the problems posed.

Of course, it is not possible for every enterprise to own a R&D department as the set up and operation costs of such laboratories are fairly high. In addition to this, the scientific team of such a department must be of high level and capacity.

A fairly good solution is the creation of a research and development department within the company, equipped with all the necessary scientific instruments for research and development purposes which can act at the same time as a control laboratory.

Finally, the evaluation of the above mentioned cases, according to my experience, leads to the conclusion that two are the best, from many aspects, choices for the most appropriate procedures to get new knowledge .

The company makes the study by itself based on the ability of its own R&D department or through a research project with an academic or private research institution especially when the project exceeds the capacity of a R&D department.

Within this context, academic institutes can play a significant role by promoting their collaboration with enterprises of the private sector.

BULGARIAN ACADEMY OF SCIENCES - SCIENTIFIC POTENTIAL AND RESEARCH PRIORITIES

NAUM YAKIMOFF
NADEJDA PETROVA
Bulgarian Academy of Sciences
15 Noemvri Str. N 1, BG-1040 - Sofia, Bulgaria

1. Introduction

In 1869, a Bulgarian Learned Society was founded in Braila, Romania. It was renamed Bulgarian Academy of Sciences (BAS) in 1911. Up to 1947 BAS was an Academy of the classical Western European type. After 1947, as in all East European countries, BAS was restructured and reorganized following the model of the Soviet Academy of Sciences.

The Law of BAS, enacted by Parliament in late 1991, restored the autonomy of the Academy and created opportunities for its transformation into a contemporary national center for scientific research, which embodies an Assembly of the Academicians and Corresponding Members of BAS. A Programme for the restructuring of BAS was elaborated and adopted by its General Assembly in early 1993. The Academy was and still is one of the very few Bulgarian institutions that adopted a programme, comprising the basic elements of a consistent science policy. At state decision-making levels, such a policy is still non-existent. The Programme determined the main research priorities of the Academy, the structural changes envisaged, the procedures and the terms of the transformation process. This Programme was completed by the end of October 1996.

The Bulgarian Academy of Sciences presently consists of 68 scientific units engaged in basic research and development in mathematics, physics, chemistry, biology, earth and engineering sciences, social sciences and humanities. The Assembly of the Academicians and Corresponding Members of BAS has three branches: (i) Natural, Mathematical and Engineering Sciences, (ii) Biological, Medical and Agricultural Sciences and (iii) Social Sciences, Humanities and Arts. The Academy incorporates some 25 specialized, economic and all-academic auxiliary units, including a publishing house, a library, archives, a computer and communications center

Ch. Proukakis and N. Katsaros (eds.),
The New Role of the Academies of Sciences in the Balkan Countries, 75–83.
© *1997 Kluwer Academic Publishers. Printed in the Netherlands.*

etc. The regular staff of the Academy, supported by the state budget subsidy amounts to some 8,950, including 3,950 scientists.

It is for the third time in the last three years that we have the opportunity to discuss problems and prospects of the Academies of Sciences of the Balkan Countries. Some two years ago in October 1994, when we celebrated the 125th Anniversary of the Bulgarian Academy of Sciences (BAS) the attention was focussed on the international scientific cooperation [2]. At the impressive international workshop that was held in Sinaia, Romania, in April 1995, the problems of "Academies in Transition" were discussed with a large participation of representatives of the Academies of the Balkan countries. We are here now, in November 1996, to talk about the role of our Academies of Sciences. This is another sign of the vivid interest toward Academies in the region.

It is quite understandable that in the process of European integration, international scientific cooperation is of vital importance. As a matter of fact, scientists are, and have always been, the best ambassadors of their countries and, in most of the cases, they have preceded politicians in establishing contacts and fostering mutual understanding.

At the end of the second millennium, we are witnessing tremendous transformations in Europe and especially in its Central and South-Eastern part. Academies of Sciences are inextricably involved in this process. It is, therefore, important to take all possible opportunities for exchanging information about ongoing processes, to discuss positive experience and failures, which unavoidably accompany the efforts of the Academies in transformation to cope with national traditions and at the same time to adapt to European conditions, whatever that could imply.

These considerations have determined our task to present in this paper the current state of the BAS at the end of a period of transformation that has been initiated by the Law of BAS, enacted by the Bulgarian Great National Assembly in October 1991.

2. History

The origins of the Bulgarian Academy of Sciences date back to the middle of the nineteenth century, when Bulgaria was under Ottoman domination and most of its intellectuals lived in exile. In 1869, a group of Bulgarian scholars met in the Danubian town of Braila, in already independent Romania, and established The Bulgarian Learned Society, the oldest national cultural and scientific institution. After the Liberation of Bulgaria nine years later the Society was transferred to Sofia where in 1911 it was changed into the Bulgarian Academy of Sciences. The Academy was dedicated to the furtherance of knowledge and its use for the general welfare. It comprised divisions of History and Philology, of Philosophy and Social Sciences, and of Natural and Mathematical Sciences.

After World War II, in the period between 1947-1949, BAS was transformed into an Academy of the Soviet type and deprived of its traditional autonomy. Enormous disproportions were introduced under the imposed ideologization of science and the slogan "Science-a productive force". The number of academic institutes and of academicians grew sharply. Bulgarian scientists became totally dependent on "those above".

The new Law of the Bulgarian Academy of Sciences, which was passed at the end of 1991, restored the autonomy of the Academy as the National research institution with the task "to participate in the development of science in accordance with the universal human values and national interests, and to contribute to the multiplying of the intellectual and material riches of the Bulgarian Nation".

3. Structure of the Bulgarian Academy of Sciences

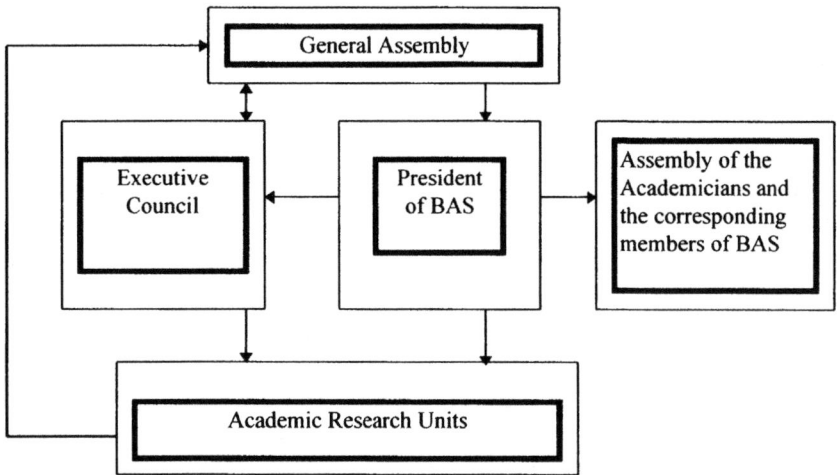

The Law of 1991 transformed the Bulgarian Academy of Sciences into a national, autonomous, non-governmental organization, combining the functions of a Learned Society and a National Center for Scientific Research. The Academy is governed by its General Assembly, its Executive Council and its President. The research units have been given a much greater autonomy within the structure of the Academy.

The General Assembly (GA) is the supreme governing body composed of senior scientists, representing the Academy's research units and elected by secret ballot. It is presided by its Chairman, two deputy chair persons and a secretary.

The General Assembly is responsible for the science policy and the general affairs of BAS. It elaborates, adopts, modifies and amends the Statutes of BAS. It is responsible for setting up and closing down the Academy's research units. It elects the President of BAS and, on his proposal, the Vice Presidents, the Scientific Secretary General and the Scientific Secretaries of BAS. It allocates the budget of the Academy and is responsible for the Academy's property, investments, etc. A number of GA standing committees are concerned with policy matters.

The Executive Council (EC) of the Academy is composed of 25 members, elected by the GA, vested with the responsibility to implement the decisions of GA. The President of the Academy, by statute, presides over the EC. The Vice-Presidents, the Scientific Secretary General and the Scientific Secretaries are members of the EC by statute. The EC elects by secret ballot the directors of the research units of the Academy. It is responsible for the assessment of the research units. It is vested with the responsibility for the Academy's international cooperation.

The President of BAS. Eligibility to presidentship is restricted to academicians and corresponding members of BAS. The President of BAS represents the Academy in the country and abroad, allocates and governs the budget of the Academy in accordance with decisions taken by the GA. He appoints the Vice-Presidents, the Scientific Secretary General and the Scientific Secretaries elected by the GA, as well as the Directors of the Research units elected by the EC. He is charged with responsibility of organizing and supervising the implementation of all academic activities. He is entrusted with the responsibility to conclude contracts and agreements on behalf of BAS. The President of BAS chairs the Assembly of the Academicians and the Corresponding Members of BAS.

The Assembly of the Academicians and the Corresponding Members of BAS (AACM) is composed of national and foreign members. It has three branches: a Branch of Natural, Mathematical and Engineering Sciences; a Branch of Biological, Medical and Agricultural Sciences and a Branch of Social Sciences, Humanities and Arts. Only academicians are entitled to elect members of the AACM. According to the Law of BAS the total number of academicians should not exceed 80 and that of the corresponding members - 120. The last elections of academicians and corresponding members were held in 1995, some six years after the preceding elections. Twelve academicians and 25 corresponding members were elected. AACM is now composed of 46 academicians and 87 corresponding members.

The research units of the Academy are legal entities performing basic and applied research, postgraduate and postdoctoral training.

4. Research Priorities and Scientific Potential of BAS

It was in April 1993 that the General Assembly adopted a program for the transformation of BAS. The Academy proved to be and still is the only scientific institution in Bulgaria, which started a reform despite the still continuing lack of a national science policy. The elaboration of the program was preceded by a preliminary assessment, which was subjected to discussion at many levels for more than a year. The program rested on three prerequisites: the restructuring should (i) be a well corroborated evolution and by no means a revolution; (ii) take into account established traditions and achievements and (iii) be based on a set of predetermined and accepted research priorities. Thus, a science policy and strategy has been worked out and made publicly known. It has been implemented during the last three years.

Research conducted at the BAS serves the following national priorities:

- Informatics, Communications and Control
- Energy Sources and Efficient Use of Energy
- New Materials and Technologies
- Environment and Environmental Protection
- Nature and Raw Material Resources of Bulgaria
- Study of Man and of Living Nature
- Exploration of the Earth
- Bulgarian National History and Culture
- Relations and Structures in Society

Ensuing from the most rational implementation of these priorities, the Academy started its structural reform. It implied the preservation of 28 research units, the closing down of 27, the establishment of 20 new ones and the reorganization of 12 units. The fate of another 19 units depended on the result of a thorough assessment of their research activities. As a result the number of the research units was diminished by 23% to 68 at the end of 1995, as compared to the 88 units at the end of 1990.

The closing down of the academic units, however, did not automatically mean the dismissal of qualified specialists, but rather the regrouping of units with a related research domain. Still a considerable reduction of about 35% of the total number of employees took place between 1990 and 1996 (from almost 15,000 to less than 9,500 employees). This included most of all inadequately extended administrative and auxiliary staff. The number of the scientists was reduced by some 900 (18.5%), from about 4,850 to 3,950. Part of these 900 scientists retired. Others were employed abroad, or joined Bulgarian commercial companies or other enterprises including economic or technological self-financing units of BAS.

The reduction of the academic institutes and staff that resulted from the transformation of the Bulgarian Academy of Sciences has had no effect on the overall research activities. The publications of the Academy within the last four years were not influenced by the ongoing restructuring. Despite the fact that only about 16% of all Bulgarian scientists are employed at the research units of the Academy, they continue to produce more than 65% of the Bulgarian publications in peer-review journals.

The research units and their staff are distributed within the eight major fields of science, covered by BAS:

1. **Mathematics and Mechanics.** Three research units, a staff of 500, including 340 scientists, perform experimental and theoretical research in mathematical structure and modelling, mathematical foundations of computer technologies and communications, informatics, computer virology, methods and algorithms for information processing, mechanics and biomechanics.

2. **Physics.** Seven research units, a staff of 1,250, including 700 scientists, perform basic and applied research in theoretical, nuclear, elementary particles, high energies and condensed matter physics, physical and quantum electronics, radiophysics, astronomy, space physics and space material science. The problems of energy, renewable energy sources, including solar energy, fluid and plasma physics are studied.

3. **Chemistry.** Eight research units, a staff of 965, including 470 scientists are involved in theoretical and experimental research on the methods of synthesis and analyses of new inorganic, organic and bioorganic substances with predetermined properties, the mechanisms of homogeneous and heterogeneous catalysis, thin foam films and colloid systems, diverse application of photoreactions in thin layers, new media for optic storage of information, electrochemical processes, methods of operating processes in engineering chemistry, new technologies of polymer production.

4. **Biology.** Fifteen research units, a staff of 1,720, including 730 scientists, fulfill basic and applied research in molecular and cell biology, neurobiology, biomedicine, the regulatory and adaptive processes in plants, animals and humans, in bioengineering and biotechnology. Well established is research on Bulgarian flora and fauna, the flora and fauna of the Black Sea and the Danube river regions. Special attention is paid to studies of ecosystems with emphasis on preservation of biodiversity and on ensuring a sustainable development of bioresourses.

5. **Earth Sciences.** Eleven research units, a staff of 1,600, including 450 scientists perform complex research in the fields of geology, geography,

geodesy, seismic mechanics and engineering, mineralogy, seismology, physics of the atmosphere and the cosmic space around the earth, meteorology, hydrology, water resources, geophysics, geotechnics, cosmic studies and remote sensing, marine sciences. The Academy institutes are involved in global and regional programs, including cosmic programs, devoted to studying the planet earth and its dynamics.

6. **Engineering Sciences.** Six research units, a staff of 940, including 415 scientists are working in the field of metals sciences and technology, the development of new materials, including composite materials. Other major activities pertain to developing new computing and communication systems, information technologies, control and system analysis, mechatronics, computer engineering and artificial intelligence.

7. **Humanities.** Eleven research units, a staff of 840, including 520 scientists are concerned with studies of Bulgarian history, language, literature, art, cultural heritage, and the Bulgarian contributions to World Civilization. Other research fields are archaeology, Balkan studies, Thracian studies, ethnography and folklore studies, Slavic literature, linguistics, the history and the theory of fine arts, theatre, cinema, architecture and music, the interaction between Bulgarian and World culture.

8. **Social Sciences.** Seven research units, a staff of 435, including 300 scientists perform studies concerned with economic, social and legal relations and structures, the history and theory of law and legal studies, the history and contemporary developments in philosophy and the studies of science, the problems of human resources and mobility, the theory and practice of sociology, demography, economics, and politology.

The Academy incorporates some 25 specialized, auxiliary, economic and technological units, including the publishing house and the library of BAS (both created 127 years ago, together with the Bulgarian Learned Society), an archive, a computer and communication centre etc.

5. International Scientific Cooperation

A priority in the policy of BAS is the development and strengthening of the international scientific cooperation. The practice of publishing in renown international scientific journals and the sharp rise in the possibilities of establishing contacts with foreign colleagues and institutions have created very good conditions for the recognition of the achievements of the Bulgarian scientists.

The Bulgarian Academy of Sciences is in bilateral partnership relations with more than 30 mostly European Academies and National scientific institutions and is a member of 22 governmental and non-governmental international scientific organizations. There is a steady trend for increasing the number of international joint projects both within those bilateral agreements and the different programs launched by international organizations such as the UN, UNESCO, ICSU, FAO, WHO and especially by the EC (PECO, INCO-COPERNICUS, EURECA, COST, TEMPUS, ESPRIT, PHARE etc.) and NATO. The number of joint publications with foreign scientists doubled in 1995 in comparison with 1990, amounting to 630 out of the some 2,300 published in peer-review journals.

The task of the further development of the international cooperation of BAS is the development of the contacts and the implementation of joint projects with the academies and the research centers in South-East Europe. There are important areas which might be explored with the efforts of the scientists in the region; let me mention the Black Sea and the Danube river, the problems of biodiversity, the environmental monitoring, the preservation of nature and cultural heritage, the authentic folklore, to name only a few of the fields of common interest.

6. Conclusion

We have presented the Bulgarian Academy of Sciences as it appears seven years after the political changes in Europe. These were years of transformation and restructuring of the Academy on solid ground, created by the new Law of BAS. Born in difficult times, the Academy has proved its right of existence. It withstood the radical destructive tendencies which were common for most of the ex-socialist countries and set up a program for the transformation of the Academy into an effective national research center. In his speech at the ceremony on the occasion of the 125th anniversary of the Bulgarian Academy of Sciences [1] the late President of the Academy, Professor Jordan Malinowski, emphasized that: "The restructuring of the Academy cannot be a single act of arbitrary setting up, closing down or transforming of research units. A profound reform of a powerful research institution, which is devoid of any intention to destroy the institution, should be introduced gradually and carefully. Otherwise there is the risk of obliterating some trends in science, of ruining valuable research facilities, and chasing away highly qualified specialists".

The World Science Report of UNESCO [2], considering science in Europe, emphasizes as positive the fact that the Academies in the ex-socialist countries of Central and South-East Europe have managed to survive. The Academies are at different stages of transformation, aimed at creating conditions for further development, but still maintain a very high level of research with the best qualified scientists of the countries. Whether and how the evolution of these institutions will follow the experience of the established European centers for scientific research remains to be seen. In the Bulgarian Academy of Sciences, at the present stage of its transformation, competence, creativity and responsibility have been preserved and we are convinced that they will remain the major features of its future development.

References

1. Malinowski J., Yakimoff N., and Nedkov P., [Eds.] (1995) 125 Years Bulgarian Academy of Sciences. Proceedings of the Scientific Conference "Science in Bulgaria, UNESCO and International Scientific Cooperation". Academic Publishing House, Sofia.

2. Kuklinski A., Kacprzynski B. (1996) Etat de la Science dans le Monde. - L'Europe Centrale. dans "Rapport Mondial sur la Science". Editions UNESCO, pp. 84-95.

REALIZING THE ECONOMIC BENEFITS OF GOOD SCIENCE

ROBERT E. ARMIT
OCRI Technology Transfer Centre
Suite 400, 340 March Road, Kanata, Ontario, K2K 2E4 Canada

1. Introduction and Purpose

Allow me to express my pleasure to be with you this week in Athens to discuss the role of the Academies of Science in the Balkans. My work is in the area of technology transfer, and it is from this perspective that I wish to contribute to the discussion. In this session, I am focusing on matters of strategy and structure that assist in fostering technology transfer, and how the role of the Academies of Science might relate to these matters. Attention is directed to research in universities and to a lesser degree in government laboratories, and technology transfer in the context of university and government research. The ideas that were included in my paper in 1994 are a part of the current discussion.

2. Prioritizing and Explaining Our Actions

Education and research in universities and government are important to the world in which we live. Industry and the nature of the economy are equally important to the world in which we live. Allow me to present you with two exercises which test our prioritizing in societal systems. Some fifteen years ago, I worked at the University of Alberta in western Canada with a brilliant Vice-President (Research) by the name of Dr. Gordin Kaplan. Speaking at a university convocation in the mid-1980's, Gordin Kaplan asked the audience the following question: "if all of us as a society had to choose between having MIT - the Massachusetts Institute of Technology - or having GM - General Motors - in our immediate midst, which one would you choose?"

The question gained considerable attention. I have raised the question to various persons and groups over the years. There is a majority who would choose MIT. One reason given is that education gives people a chance to do a lot of things. A second point is that education and research offer a broader beachhead for the community. Thirdly, a school like MIT conducts excellent

85

Ch. Proukakis and N. Katsaros (eds.),
The New Role of the Academies of Sciences in the Balkan Countries, 85–95.
© *1997 Kluwer Academic Publishers. Printed in the Netherlands.*

research which has wide appeal in industry, providing its own impetus to industry. Things are changing in industry, however. Many large companies like GM are internalizing major programs of education and research. The economic multiplier effect of a large company in a geographic area creates its own broad beachhead. And few societies can afford a MIT without having a GM or strong economy.

Here is an equally challenging question asked by Joe Spina, a perceptive member of the legislature of the Province of Ontario in central Canada. Joe Spina asked a group of business persons in Ottawa the following question: "if our society had to choose between having one company employing 3.000 persons or 1.000 companies each employing 3 persons, which option would you choose?"

In this particular case, Joe Spina speaks for small business in the province and finds the 1000 company option attractive. Many agree for various reasons. But there is no right answer. Few societies can afford the 1000 small firm option without the ability of many of the small firms to cooperate in competitive bidding. This is one reason why firms grow large, they successfully internalize the cooperative mode. A strong argument exists for supporting large companies in an area. Economic development programs generally testify to the strength of the argument. Increasingly, systems that support smaller enterprise at the base are gaining attention. The new economy is characterized by a relatively larger number of small businesses.

The exercises show that we as a society value education and research institutes and small and medium sized enterprise not only for themselves but also in a relative context. The first exercise also indicates the importance of involvement of education and research in the community and how industry relates to this education and research. This is very much technology transfer.

3. Good Science and Economic Benefits

Good science is defined for purposes of this paper as scientific research which contributes to the world around us in a manner of substance. Research enhances understanding of things as they essentially are and advances knowledge in a cumulative manner. Implicit in the definition of good science is a sharing of the results of research with others. This sharing ultimately contributes through education or through the economy or both.

Scientific work reaches out and is shared in various ways. Major findings are circulated rapidly world-wide scientist to scientist. Publications are an important vehicle for the transfer of ideas and where science is moving. Education programs reflect the current state of knowledge in specific fields. It is worth noting that the patent literature also contributes to the directions of science, and that industrial research is similarly an important part of the literature.

Economic benefit from good science represents incremental economic gain in the system that comes from the application of science. One source of science is university and government research. At the interface of university and government based science and the economy, there are several vehicles of technology transfer that are now taking hold. These include the following:

1. Publications, circulars, journals and reports
2. Consulting engagements
3. Conferences, seminars and workshops
4. Institutes, centres and groups
5. Affiliate programs
6. Research consortia
7. Patents and licenses
8. Contract engagements
9. Facility and equipment arrangements
10. Research and science parks and new business incubators
11. Spin-off company formation
12. Joint ventures
13. Guest company arrangements
14. Personnel exchanges
15. Students, undergraduates and graduates
16. Continuing education programs

Some of you will recognize this list from my presentation in 1994. The report from the meeting has a discussion of the various vehicles of technology transfer. Industry increasingly searches out the ideas generated in the scientific laboratories in universities and government. There is a heightened relationship between science and the economy, almost as strong as the tie of scientist to scientist. More and more universities and government laboratories are looking to the contribution of their science to the economy through the various vehicles of technology transfer. This is part of the culture of the "new" economy. The organizations in the "new" economy place importance on new ideas, innovation, research and development, new product introduction, shorter product life cycles, global markets, interorganizational relationships, industry/university relationships and industry/government collaboration in new forms, careers with a tie to continuing education and networking among persons and organizations.

4. Strategy of Technology Transfer

Technology transfer is required for science emanating from the universities and government laboratories to have economic benefit. Science in this context covers areas of the natural sciences, engineering, medical science, pharmaceutical science and agricultural science. These areas have ready parallels in the economy and in industrial practice. One strategy in technology transfer follows as it might fit with a university or government research centre:

- Measure out the current situation. This is the starting point. Look at the nature of the research in the laboratory and the professional and technical staff and equipment involved.

- Recognize the various forms and vehicles of technology transfer. The list provided in this paper serves as a starting point. Consider each of these forms as they might relate to the target technology.

- Use them and adapt them as they apply to different situations. For example, has a patent been looked at in the research? Does the research lend itself to a bench scale-up and transfer to industry through a research consortia? Are there lateral applications for the technology coming out of the research which may draw attention from different industrial groups?

- Sequence and combine the forms of technology transfer effectively. For example, it is valuable to consider the patent option before there is a public disclosure of the research in a publication or seminar. Often training can be built into a contract for research or for technology transfer.

5. Structure of Technology Transfer

Structure represents the more or less ongoing patterns of relationships that exist. Structure can relate to relationships among persons and among organizations. In technology transfer major structural issues include the ways in which technology transfer is played out; the relationships between the university and government laboratories on one side as senders of technology and industry and the economy on the other side as receivers of technology; and the intermediary organizations that relate to each side of the sender/receiver relationship or which are instrumental in matters of awareness, action, and standards in the field. In Canada, there are now a plethora of organizations that include some dimension of technology transfer as part of their function. Here are several of these organizations:

1. AUTM, the Association of University Technology Managers. This organization grew out of the Society of University Patent Administrators in the United States and Canada. Many members are now from industry. The annual meeting draws 600 persons from all sectors, and regional

meetings are prominent. An excellent handbook on technology transfer has been put together by AUTM.

2. LES, the Licensing Executive Society. This is the parallel in the patent and licensing field to AUTM. It is an international organization with over one thousand members, and a strong reputation in the field. LES has a number of licensing books and an excellent magazine.

3. CAURA, the Canadian Association of University Research Administrators. CAURA includes all of the Canadian universities. The annual meeting allows members to compare notes on research policy matters, sponsored research trends and technology transfer.

4. AURRP, the Association of University Related Research Parks. AURRP is based in the United States but includes a majority of Canadian research parks and many research parks from throughout the world. The annual meeting addresses research park issues in university development, property issues and technology transfer, including areas like spin-off companies and relationships with research institutes.

5. IASP, the International Association of Science Parks. IASP is headquartered in Europe and covers science park development in various parts of the world through chapters. Their meetings draw an international representation, allowing park managers, companies and officials involved with parks to compare parks and ideas. Joint meetings with AURRP are held periodically.

6. NBIA, the National Business Incubation Association. This organization in the United States grew out of the attention in the last twenty years to "incubating" new enterprise though mentoring systems and a range of services to assist entrepreneurs in their formative years. Technology business incubators are a subset of the field, and important to university based technology transfer programs. NBIA has an open membership, superb publications and member services.

7. CHEF, the Corporate Higher Education Forum. CHEF represents a meeting place for Presidents of large companies and large universities in Canada. Their goal is wider common understanding and activity. CHEF took a leadership role in enhancing university/industry relationships, conducted a number of important studies in the field and annually makes two prestigious awards at the university/industry interface.

8. CB, the Conference Board of Canada. The Conference Board has a research committee which among other subjects considers topics in

technology transfer, training and networking. The ties to the Conference Board in the United States is also valuable. Members are companies.

9. TTS, the Technology Transfer Society. The Technology Transfer Society is an organization of individuals set up in the United States to advance technology transfer. TTS has a circular publication and annual meeting on different themes. The federal US research laboratories are well represented in the TTS.

10. NCE, the National Centers of Excellence. Canada has a program supporting networks of centres of excellence in sixteen disciplines ranging from protein research to telecommunications. Centers typically have universities, industry and government involvement. Each has a host and series of nodes across Canada. Issues of technology transfer are internalized within the network, and external possibilities are considered as well.

11. CNRC/IRAP, the Canada National Research Council and its Industrial Research Assistance Program. IRAP has received international praise for its unique brand of support for early stage industrial research. Many projects come out of university laboratories.

12. CIIC, the Canadian Industrial Innovation Centre. CIIC is set up as a not-for -profit company to assist in all forms of innovation in Canada, but particularly inventors and persons seeking advice on intellectual property protection. Universities work well with the CIIC.

13. FDB, the Federal Development Bank. The FDB funds ventures in industry across Canada and provides an array of courses geared to small business creation and business planning.

14. CATA, the Canadian Advanced Technology Association. CATA is a members organization dedicated to enhancing advanced technology enterprise in Canada. CATA comments on tax policy in the country, has meetings geared to members needs and offers members services like travel assistance and insurance programs, of considerable value to small and medium sized enterprise.

In addition to these associations and organizations, there are a number of related groups that in one way or another address technology transfer. For example, the venture capital companies have an association that accepts university based spin-off companies for its inventor forums. Manufacturers in Canada have a strong association in their field that is open to technology initiatives. There is a lot of activity in this field, and no exclusives to any group.

6. Technology Entrepreneurship and the Venture Capitalist

Money is a problem anywhere you go in the world. We recently hosted a meeting in Ottawa on spin-off companies. One of the feature subject sessions involved the technology entrepreneur and the venture capitalist. Harry Davis from the University of Victoria presented a unique view of the two areas. A chart has been prepared to reflect his idea. Essentially, Davis argues that there are four areas of attention in a technology company: technology, finances, the market and management. A technology entrepreneur views the priority of the four areas in the order presented: technology is first, and foremost, the money to support the technology is second, the market is third and the fact that we have to manage the enterprise is accepted but is least important when cast against the other three areas. A venture capitalist sees the picture quite differently. In fact, the venture capitalist sees the four functional areas in the reverse order. The most important areas to the venture capitalist is management, the ability to keep things together, spend wisely and handle well all resources. This is followed by a market for the product, a vector for production. Finances are third, including risk and spreading risk among sources of finances. Fourth is the matter of the technology itself and the science in focus.

If this surmise is true, and it had good uptake at the meeting and following the session, than is there any wonder that technology entrepreneurs run their enterprise differently than do venture firms and professional managers? And that the dialogue between technology entrepreneurs and venture capitalists is slow in starting and delicate in process? Different professional groups approach the world and the economy differently. I am also reminded of a statement made by Dr. Jai Nigam from India that there is a dire need for professional financial management in research in the world. This represents a merger between the ideas and priorities of the technology entrepreneur on one side and the venture capitalist on the other.

7. The OCRI Story

My involvement with technology transfer is through the OCRI, the Ottawa Carleton Research Institute. This is a member's organization formed in 1984 in Ottawa, the capital city of Canada and home of a substantive research effort in all sectors of industry, government and the universities. OCRI is designed to increase interaction among persons from these sectors, increase resources available for research and promote the development of the region in high technology. A good part of what is undertaken surrounds the members. Highly successful technology executive breakfasts are held monthly. Special initiatives around the electronic highway and research park development are undertaken. Training courses are offered from a number of platforms and organizations. A number of university chair programs were initiated in OCRI and continue to have a base in OCRI. This is regional force directed to the interaction among the educational institutions, government research laboratories and industry, supported by these organizations.

8. The Role of the Academies of Science

I want to bring some of these issues and situations outlined together in terms of the seminar theme. The Academies of Science in countries provide a networking organization for scientists and a voice for science. Does an Academy of Science have a role in matters of technology transfer and of realizing the economic benefits of science? And if there is a role in this respect, what types of organizational models are there?

As a proposition, the Academy or members of the Academy may choose to become more involved in matters economic while retaining the focus of the Academy in its original form. One way to become more involved is through a new arm of the existing organization dedicated to work in science and the economy. Many organizations grow by adding on new divisions or portfolios of activity. A second option is a separately chartered spin-off organization of the Academy, created solely by the Academy or jointly with other organizations including industrial associations. A third option is for scientists and organizations within the Academy who support technology transfer to establish a new organization which is itself removed from the Academy of Science and is committed to doing more with science in terms of realizing its economic benefits.

This discussion can conclude in different positions in different countries and in different time periods. From my outline of the strategy and structure of technology transfer in Canada, the end point in Canada is pretty clear. The structure of technology transfer involves a number of organizations all somewhat specialized in their own way. There is no one centre of gravity. Drawing from the example of the differences in approach between the technology entrepreneur and the venture capitalist, there may be a danger in being lead too much by the technologist. Accommodations to the business approach are important and are best represented by the business person. Here are some guidelines as I see them for technology transfer to work with the Academy involved:

1. All specialized areas are best covered off in separate efforts by specialists. This means finances involve money persons, marketing involves marketers and management involves managers. I am not arguing that scientists cannot manage; some of the best managers are scientists. As a

rule, however, professionals in different fields bring different skills to the table and building on these skills is a strength. Whichever way an Academy chooses to go, then, it is wise to involve others in the new work.

2. Networks are prominent. Networks involve persons, contacts, relationships and influencers. Networking allows many organizations, some highly specialized, some less so, to be successful. The Academy might adopt a strategy in networking for members and for science generally in their areas. Where it makes sense for new types of networks or alliances to be formed, the Academy may play a lead role in the new work.

3. Management is a focus. Patience, resources including money, goals and a vision are needed. Technology companies take time to form, research takes time to scale up, markets take time to unfold, specialists take time to work together, organizations require openness for cooperative efforts, scientists take time to move beyond their science.

4. Maintaining a strong voice for science is basic. There is a frustration in science among scientists when the results of their science are not being employed. This frustration can often lead scientists to want to undertake more work in the application of science and in technology transfer. Care must be exercised that this new work is not at the sacrifice of scientists maintaining a strong voice for science.

I attended a meeting of WAITRO in Sofia in 1993. WAITRO is the World Association of Industrial and Technical Research Organizations. The focus was the move to market economies in the eastern European arena. Dr. Nikolai Mateev and Dr. Georgi Kalushev, both from Bulgaria, spoke of the support of the Academy of Sciences for science and the pressure on the Academy to redesign in the wake of the declines in employment in the research and development organizations in their country as part of the transition. Dr. Laszlo Zsinka from Hungary informed us at the same meeting that the Hungarian Academy of Sciences had taken a 51% interest in an isotope production facility which had been formed as a spin-off company in the dividing of the Institute of Isotopes into a production company and a research institute. This research institute stayed under the wing of the Hungarian Academy of Sciences. These are significant situations. I similarly have vivid recall of a comparison drawn between MIT and the University LETI in Leningrad, now St. Petersburg, by Dr. Oleg Alekseyev, its Rector in 1991. The comparison showed that the key differences between these organizations were in resources and notably in currency, dollars and rubles. Scientific output measures were remarkably similar in magnitude. Significantly, the vectors for scientific development and movement into their

94

respective economies had parallels. There is a great deal happening in eastern Europe right now.

Education and the economy are not always in balance. Similarly, research in universities and government and the economy and systems of technology transfer are not always in balance or equilibrium. One area or system can be ahead or behind of the other. For example, what occurs when there is no ready industrial receptor for the technology coming out of the scientific laboratory in the university or government? Ideally, this is a case where the scientist might be able to set up a spin-off company, and bring all of the forces of business to roost on this opportunity. Perhaps if there is no delivery system, the opportunity is foregone. Similarly, what occurs when an industry needs work in a specific discipline of science and no expertise exists in the education and government system to provide the service. These are subjects the Academy may choose to address. It is valuable to look at how the technology sector of the economy and connections between industry and university and government research are currently performing, how balances exist and how the systems can be improved. It is a field looked at in all industrialized countries.

The Academies of Science in eastern Europe have grown up though times that are different than today. Strategies and structures are going to change. As the roles of the various Academies change, the way in which each particular Academy works with other groups in society as a whole becomes more important. These groups are also changing. Scientists add value to science and to other parts of the community including the economy. Technology transfer is a field ripe for an involvement from the Academy. My sense is that this is best undertaken cooperatively with other groups in your wider community. In Canada, we have many organizations with ties to many other organizations to achieve multidisciplinary objectives. Part of your challenge in science involves ties to management, market and finance specialists. It also focuses on better linkages between education and the economy. There are other groups in your immediate and extended societies who can relate to your challenge of doing more with science and ensuring that science and scientists survive in the transition to market economy and the "new" economy. The leadership in bringing these forces together may well be part of the role of the Academy. We wish you well in your important work.

November, 1996

NOTES:

This paper was prepared for presentation at the NATO Advanced Research Workshop in Athens, Greece in the period November 19 to 23, 1996. The subject of the Workshop is "The New Role of the Academies of Sciences in the Balkan Countries".

Robert Armit can be reached at the Ottawa Carleton Research Institute in Ottawa Canada : by telephone 1 613 592 8160; by facsimile 1 613 592 8163; and by email: barmit@ocri.ca

The WAITRO meeting (1993) referred to in the paper has been summarized as a publication available from the WAITRO office in the Danish Technological Institute Technology Park, Aarhus, Denmark which can be contacted by facsimile at the number: +45 89 43 89 89. The publication is titled "Restructuring and Management of Industrial and Technological R&D Institutions in Eastern Europe in the Transition to Market Economy".

Proceedings of the meeting of the Technology Transfer Society on Internationalizing Technology Transfer (1992) are published by the Society in the United States.

Support for the participation of Robert Armit at the Athens workshop has been provided by the workshop organizing committee, by NATO through the organizing committee and by the Ottawa Carleton Research Institute in Ottawa, Canada. This support is appreciated.

TURKISH ACADEMY OF SCIENCES (TÜBA)

PROF. DR. AYHAN O. ÇAVDAR[*]
President, Turkish Academy of Sciences
Atatürk Bulvari No 221, Kavaklidere, Ankara - Turkey

The Turkish Academy of Sciences attaches utmost importance to scientific cooperation among countries, because it believes that it not only paves the way to a constructive dialogue among different cultures, but also acts as a powerful antidote to intolerance and to ideological and racial barriers and that universality, freedom and critical thinking constitute basic elements in the scientific process and form a common bond between all cultures.

Science and technology are advancing at an ever increasing pace in our time. This fact brings us face to face with new problems.
What will the new technologies bring to our life?
Will our societies be able to absorb and digest these innovations in their stride?

It is increasingly believed that, in the 21st century, educational institutions will have to produce young people with more adaptability, versatility and resilience, who can accommodate innovations more easily than in the past and who can change their line of work easily when necessary.

Education and training will become a life-long matter for adults as well as for young people. This will mean more and better schools with more diversified programs, more and better equipped teachers, new institutions to form and train such teachers, greater efforts to produce gifted scientists and researchers. Educational costs will therefore rise up and will cut deep into the budgets of governments. These are some of the challenges which will confront governments in the 21st century.

It is obvious that it will be the duty of our academies to tackle some of these problems. As Balkan countries, our duties will be all the more arduous, as we will have to cope with inadequate resources.

Our Academy has taken up this matter with our Ministry of Education and has recommended setting up a mixed commission to revise the whole educational system. There are other things that are incumbent upon us.

---

[*] given by Prof. Dr. Ayhan Ulubelen

Ch. Proukakis and N. Katsaros (eds.),
The New Role of the Academies of Sciences in the Balkan Countries, 97–105.
© *1997 Kluwer Academic Publishers. Printed in the Netherlands.*

Recently, applied research seems preferred to basic research. Of course, applied research is very important for the economic development of the countries, but without basic research the technical capacities will be seriously weakened. In the long run, applied research needs a strong basic research.

Due to the lack of sufficient funds and of a scientific approach among the leaders of the countries, basic research is considered a waste of time and money. This is especially so in developing countries where there are a number of problems to be solved quickly and not enough funds and trained man-power.

Disparities between poor and rich nations are very high. Ever growing production of goods and the number of new technologies are numerous; therefore, the developing nations do not want to lose more time doing basic research, but try and catch up with the developed countries. After the Second World War, for a short time, some Asian countries and Japan did almost no basic research. After reaching a certain economic level, the Japanese government encouraged work on basic research, for in a society where basic research activity is too weak and it is considered that the natural sciences have no part in their culture and the only interest is to be powerful economically and militarily, this will put out the creative man power from the field of science and in the long run these countries will start losing the ground.

In Turkey we are trying to balance the support of basic and applied research. In basic research, we have three different kinds of organizational support:

1. The University Research Funds, support the scientists on a project basis, in some universities these funds work very well and many young and senior scientists get rather handsome support. But this does not (usually) cover infrastructure. For the infrastructure, the State Planning Organization gives support on a project basis. This, of course, may not always be enough.

2. The second Organization which supports the basic sciences in Turkey is TÜBÝTAK (The Scientific and Technical Research Council of Turkey).

    They give support to basic and health sciences as well as engineering sciences on a project basis. One can get small scientific instruments sometimes with a worth of up to 10.000 dollars as well as money for travel and for reagents etc.

    TÜBÝTAK also supports integrated doctorate programs. In recent years, they initiated a program to support industrial research, especially for small and medium size industries.

TÜBÝTAK is establishing a Technopark for these groups of industry; another Technopark is already established at the Middle East Technical University in Ankara (ODTÜ). Another organization for the support of small and medium size industry is TTGV which is the Technology Development Foundation of Turkey.

Private big industry is collaborating with western countries more and more. It used to buy know-how for many years, now this tendency is slowly changing to run their own research and develop new products themselves. Turkey is also collaborating with a number of developing countries giving them know-how, supporting them financially and building some necessary industrial plants.

3.   The third organization to support basic sciences is TÜBA. So let us go back to how the Turkish Academy of Sciences was founded, and what are its activities.

1. The Founding of the Turkish Academy of Sciences

The Turkish Academy of Sciences (TÜBA) was founded by decree-law No:497 on September 2, 1993.

2. TÜBA's Aims and Objectives

TÜBA's main aims and objectives are as follows:

To stimulate scientific curiosity. To awaken interest in research and science. To confer awards on successful scientists for their creative work as well as on potential scientists for encouragement in this domain. To spread scientific methods and thinking, to upgrade the social status and prestige of scientists and researchers, their standard of living and the special prerogatives which their activities require. To take the necessary steps with the government to materialize these aims and objectives.

TÜBA is a corporate body with scientific, administrative and financial autonomy. It is attached to the Prime Minister's office. (The Prime Minister uses his or her "authority" over TÜBA when necessary, through a cabinet minister of his choice). TÜBA's activities are governed by private law, except those specified in article 2 of the decree-law relating to its creation.

3. The Organization of TÜBA

According to the relevant article of decree-law relating to TÜBA's foundation, organs of the Academy are:

- The General Assembly of the Academy
- The Council of the Academy
- The President of the Academy

3.1. THE GENERAL ASSEMBLY

The General Assembly is composed of the Academy members. It meets at least once a year, at the time and place assigned by the Academy council. The functions of the general assembly as determined by article 8 of the decree-law, are as follows:

To debate and to ratify the activity report, the balance sheet, the draft budget and the fundamental documents relating to science policy prepared by the Academy Council.

To elect new academy members.

To elect new members to replace the outgoing members of the Academy Council.

3.2. THE ACADEMY COUNCIL

The Academy council consists of 10 members and the President of the Academy, who is elected for four years by the General Assembly. Five of the members of the council lose their membership every two years by drawing lots. They are replaced by new members elected by the general assembly.

The functions of the Academy council, as specified in article 10 of the decree-law, are as follows:

- To program and carry out the activities of the Academy.
- To set up commissions within or outside the Academy for counseling or studying purposes.
- To prepare the draft budget.

3.3. THE PRESIDENT OF THE ACADEMY

According to article 11 of the decree-law relating to the foundation of TÜBA, the president of the Academy is elected by the General Assembly among its members and appointed by the Prime Minister for four years.

The president of the Academy is responsible for keeping the Academy's activities within the scope of its aims and objectives. Payments on behalf of the Academy come within the president's jurisdiction.

4. Foundation Stages of TÜBA

In accordance with the provisional article 1 of the decree-law Nr. 497 pertaining to the foundation of the Academy, and in compliance with the proposal of the Scientific and Technical Research Council of Turkey (TÜBÝTAK), 10 members of the Academy were chosen by the Prime Minister once for all, upon the foundation of TÜBA on September 2, 1993. These 10 members, chose, in their turn, again in accordance with the same provisional article 1, 10 other members, completing in this way, the constitution of the general assembly.

The general assembly held its first meeting on November 27, 1993, determined the members of the Academy Council and elected Prof. Dr. O. Çavdar, as the first president of the Academy. The organs of the Academy have thus been completed.

The Academy council began its activities on January 7, 1994, holding its first meeting on that day. The general assembly met 8 times and the council 34 times by October 1996 inclusive.

5. Members of The Academy

TÜBA has three kinds of members:
- Principal Members
- Associate Members
- Honorary Members

As has been stated in article 3 of the decree-law Nr. 497 pertaining to its foundation, the number of "principal members" has been restricted to 2% of all the professors in universities in Turkey.

As for the "associate members", their number cannot exceed the threefold of the principal members' total.

There isn't any numerical restriction for "honorary members".

TÜBA has now 93 "members" elected and approved by the general assembly upon the proposal of the council. The elections are made with utmost care and vigilance.

The number of TÜBA members working in various domains are shown below:

Scientific domain	Honorary Members	Principal Members	Associate Members	Total
Medical Sciences	3	10	6	19
Natural and Engineering Sciences	7	23	21	51
Social Sciences	11	9	3	23
Total	21	42	30	93

6. Committees

The council decided to set up committees for them to pursue their scientific activities in their respective domains and develop proposals. The different committees thus set up and their scientific domains are as follows:

- Committee for the Science Policy
- Committee for the Scientific Meetings
- Committee for Sciences and Ethics
- Committee for Relations with Foreign Countries
- Committee for Seeking and Proposing Candidates
- Computers and Network Committee
- Human Rights Committee

7. Programs for Motivating Purposes in Social Sciences

The Turkish Scientific and Technical Research Council (TÜBÝTAK) is encouraging research activities in the basic and applied fields of natural, engineering, health and agricultural sciences. There is no institution in Turkey, as yet, to render this service in the domain of social sciences.

- TÜBA has decided to fill this gap in a certain measure, by taking interest in the praiseworthy activities and research work of social scientists who carry Turkish citizenship, by noting their high qualifications, bringing them to public notice and by granting them awards to encourage them until a "Social and Economic Research Council" has been founded, or until these duties have been assumed by TÜBÝTAK.

  TÜBA has instituted for that purpose three kinds of awards, namely: science awards, service awards, and encouragement awards.

  According to the principles which have been adopted, the science awards are granted to scientists who are still living and who have made important contributions to science, at an international level, by their scientific activities.

  The service awards are granted for the training of scientists, for the institutionalizing of scientific fields and for important contributions to the development of social sciences in our country.

  The encouragement awards are granted to gifted, promising young researchers with potential to make contributions to science at an international level in the future and who are not over 40 years on the first day of the year in which the award is granted.

- TÜBA has also instituted scholarships at the doctorate and post-doctorate level to encourage studies in the field of social sciences and has earmarked 8 billion TL. in its 1996 budget for that purpose. The rating of the applications will be made by a commission set up for that purpose. These scholarships are also intended to finance research activities abroad in an effort to enable researchers to continue their work at international research centers.

- TÜBA worked out a program in order to motivate all researchers and teaching staff of Turkish citizenship working in the research centers of the public and private sector, into issuing publications at an international level in the domain of social sciences.
  In this respect, articles at an international level, published in journals which have been judged noteworthy by "Social Sciences Citation Index", "Science Citation Index" or "Arts and Humanities Index", as well as selections, original works and excerpts from selections published by well known publishers in developed countries are considered as publications at an international level.
  To motivate scientific research, our Academy grants awards within the framework of this program, for publications falling into the above-mentioned category.

- The Turkish Academy of Sciences has adopted some guiding principles for the support of scientific periodicals published in Turkey. Priority is given in this respect to periodicals in the field of social sciences. The support accorded can be financial as well as scientific approval.

- TÜBA also supports international scientific meetings in the field of social sciences. The aim followed in this respect is to increase international scientific cooperation raising thereby Turkey's own level in social sciences.

8. Scientific Meetings

TÜBA has arranged a series of conferences to keep science in the public eye, or -better said- to make it a part of daily life. Two of these meetings were held in 1994, their subjects being "Science, Ethics and the University in the World and in Turkey" and "Science and Education" respectively. The books about these two meetings have already been published.

As for 1995, TÜBA decided to hold a series of conferences and panels under the general heading "University". In that year three meetings were convened in compliance with that decision on the subjects: "The University Statute" on April 28, 1995, "Academical Promotions in Universities" on June

23, 1995; "What Kind of a University Graduate Do We Want?" on November 24, 1995.

9. TÜBA Publications

Five books have been published which cover the panel meetings, discussions and the communiqués of TÜBA's scientific meetings. The first book was about "Science, Ethics and the University", the second book was related to "Science and Education", the third book, relating to the first meeting covered the subject of the "University Statute", the fourth book entitled "Academical Promotions at Universities" and lastly the fifth was about "What Kind of a University Graduate Do We Want?".

10. Conferences at Universities

According to Article 4 of the decree-law Nr.497, relating to the foundation of TÜBA, very high scientific qualifications are required in the election of members. Only the most qualified and distinguished scientists are, therefore, to be found in the cadres of TÜBA. TÜBA prepared a program of scientific conferences to use this vast pool of qualified scientists for the benefit of the newly founded universities throughout the country.

11. Diffusion of Scientific Thoughts

The Turkish Academy of Sciences has adopted the principle of discussing with an open mind all problems relating to science and scientific thought, of publicizing the results of these discussions and of diffusing its views.

It has, so far, publicized its views on:

• Freedom of Thought
• Autonomy of the Scientific Organizations
• Science and the Future of Turkey

12. Relations with Other Countries

TÜBA President Prof. Dr. Ayhan O. Çavdar and member of the Academy Council Prof. Dr. Erdoğan Þuhubi participated in the meeting of "ALL EUROPEAN ACADEMIES" (ALLEA) where they were able to establish communication with the Academies of Eastern and Western European countries which attended the meetings.

A cooperation protocol was signed with the French Academy of Sciences on March 24, 1994, providing for cooperation at the post-doctorate level in all domains of science.

Prof. Dr. Ayhan O. Çavdar and member of the Academy Council Prof.Dr. Ayhan Ulubelen took part in the second ALLEA meeting in Budapest, March 21-22, 1996.

The Bulgarian Academy of Sciences held a meeting on October 11-12, 1994, to commemorate its foundation. Prof.Dr. Ýzzet Berkel, member of the Academy Council, attended the meeting and presented a paper.

Prof. Dr.Hamit Fiþek, member of the Academy Council, attends the European Science Foundation Standing Committee for Social Sciences meetings.

President of the Dagistan Academy of Sciences Prof.Dr.Hamit Buçayev visited TÜBA president Prof.Dr.Ayhan O. Çavdar twice, on April 5, 1995, and on September 11, 1995 respectively. During the latter visit, he presented Prof.Dr.Ayhan O. Çavdar with a letter informing her that she has been admitted as member to the Dagýstan Academy of Sciences.

TÜBA sent out in August 1995 letters to 31 science academies abroad "URGING THEM TO APPEAL TO THEIR GOVERNMENTS TO TAKE ALL NECESSARY MEASURES TO STOP TORTURE, MURDER AND GENOCIDE IN BOSNIA".

TÜBA has become a member of the INTERNATIONAL HUMAN RIGHTS NETWORK OF ACADEMIES AND SCHOLARLY SOCIETIES and is taking part in its activities.

After this short expose about the aims of the Turkish Academy of Sciences, I would like to come back to the scientific collaboration among our countries and make the following suggestions as a way and means of raising this collaboration to a concrete level.

1. Our countries promote and facilitate visits of candidates of at least post doctoral or equivalent status in the natural, technological, social sciences and humanities.
2. Collaborate in research and exchange of experiences with partner colleagues.
3. Participate in joint research projects to be carried out on the basis of specific understanding between scientific organizations.
4. Participate in seminars, symposia and bilateral discussions of themes of mutual interest and exchange views and experiences on science policy.
5. Each country will endeavour to make its publications available to other countries.
6. Exchange of correspondence between delegations to evaluate the program and development of scientific cooperation when necessary.

FOR A NEW CONCEPTION AND STRONGER ROLE OF THE ACADEMY OF SCIENCES AS A PRODUCTIVE, COMPETITIVE AND COLLABORATION PROMOTING RESEARCH CENTER

P.D.SKENDE
Academy of Sciences
Square "Fan Noli", Tirana, Albania

1. Science, research and development as well as their institutional structures, have continuously been a point of discussion for the Democratic States of a free market economy and with much more unknown parameters for newcomers to this economy, those which 5-6 years ago belonged to a centrally planned economy and a totalitarian political system.

Academies of sciences as scientific institutions, during the last years, have been both object and subject of these discussions and Albania is not an exception in this respect.

In line with the subject of our Workshop, I have to mention something which I need for reference in the following presentation: talking about the role of Academies of Sciences, the semantics of the terminology is not the same or equal for all of us, even within a limited region like the Balkans.

I would like also to mention the excellent overview given by Prof. Theocaris P.S. of Athens Academy [1], the oldest one, established by Plato, which gives a picture of history, changes and evolution of Academies in all its spectrum.

Without too much rhetoric, I think that talking about a new role of Academies (somewhere also Academies of Science and Arts) for our region, we have in mind two cases:

a) Academies which are centers, gathering, according to their professionalism and contribution, the best intellectuals of the country, a kind of Pantheon and so, basically, are honorific institutions of high prestige.

b) Academies which, at the same time, include in their structure research centers with adequate scientific institutes or labs of national importance, with a high capability of expertise - theoretical or experimental - toward the needs of the country and science progress in general.

Ch. Proukakis and N. Katsaros (eds.),
The New Role of the Academies of Sciences in the Balkan Countries, 107–112.
© 1997 *Kluwer Academic Publishers. Printed in the Netherlands.*

Once more, within such classification, we have at least two levels of their role - national and regional.

The broad spectrum of institutional science management of today - national centers and national labs, universities and specialized agencies, private institutes and research units etc., clearly makes it difficult to undertake an effort to find an optimized model valid for all countries without taking into consideration the peculiarities of the socio-economic framework, historical background and the impact that these forms offered to their respective countries. In the case of Albania, one thing is clear: paradigm about the scientist's role and integration into the democratic world community as a free thinker and one of the driving forces of progress, after the terrible self isolation of half a century, must be changed and is changing rapidly. At the same time, personally, I agree with what Thomas Kuhn has written in one of his books: "The determination of shared paradigms is not, however, the determination of shared rules" [2]. And these "rules" in our discussion represent the experience of each country.

2. In the case of the Albanian Academy of Sciences, I will give a brief description of the situation at the national level. The Albanian Academy is the youngest in Europe - less than 25 years of existence - and despite this, we can call it, let's say, a model of the second type which functions both as a research center and honorific part. There were some peculiarities coming from the isolation policy more than the originality of such a model. Three years ago, a new approach was adopted. The Academy represents a system with two, clearly different, bodies:

a)   the Research Center

b)   the Assembly, a Honorific Part with weak institutional links and without important decision-making power toward the research center.

The last three to four years have been a period of endeavor to reorganize just the active part, the research center, according to the demands of a new economic order of a free market economy, with much more independence and incentive, steps for stimulating the institutes to be involved in programs and projects of R&D with interest for the country. Institutes are stimulated to be open for collaboration with foreign institutions and scientists, fulfilling, as fast as possible, the vacuum created from the isolation of the past. I can mention here, according to data of UNDP Human Development Report (Albania 1996)[3] that only during 1994 and 1995 more than 100 (from about 250) researchers of different institutes contributed to more than fifty international conferences by submitting their papers. These figures are still modest, much more compared to ten years ago.

The Research Center of the Academy is financially supported by the State and does not interfere with the decisions for the fields of activities chosen by institutes. State support is coming also by legislation which stipulates involvement in projects, national or international and allows (after 30% taxation) the use of the surplus created, according to the needs of institutes, including the possibility to pay extra fees to their staff who demonstrate motivation in a project's formulation and implementation.

The Research Center has twelve institutes, a library and publishing unit and is managed by a Board in which Directors of Institutes, a Chairman, two Vice-Chairmen and a Scientific Secretary participate. The members of the Board must have high scientific degrees, be active scientists and skilled science managers. Besides these, no other mandatory conditions related to their membership to the Assembly are required, with the exception of the Chairman.

The Board formulates the general strategy for R&D of the Center taking into account the country's priorities for social and technological progress. It supervises the budgetary distribution policy within the financial situation and publishing priorities. From the institutes, seven are of natural and technical sciences:

1. Institute of Informatics and Applied Mathematics
2. Institute of Nuclear Physics
3. Institute of Biology
4. Institute of Seismology
5. Institute of Hydrometeorology
6. Geographic Center Studies
7. National Hydraulic Lab

Five other institutes are from humanities or social sciences:

1. Institute of Linguistics and Literature
2. Institute of History
3. Institute of Archaeology
4. Institute of Folk Culture
5. Center for the Studies of Art

Some Institutes of Natural Sciences are covering also nation-wide specialized services like earthquake monitoring, collection and elaboration of hydrometeorological data including hydro, wind and solar energy potential of the country, radiation environmental monitoring etc.

3. At present, according to the possibilities and needs of the country, the main activities are focused on applied research and development with an increase of participation in scientific projects with foreign partners and international organizations. Some of them are EU projects, IAEA, IHMO, IGA, UNEP, World-Lab., NATO Scientific Division, many universities from Greece, Germany, France, Italy etc., South-East European Association (language, history) and some foreign Academies. I would like to stress that at the research center of the Academy, independently from the label, we are acting according to the demand of a country going closer and closer to the rules of market economy which must demonstrate competitivity in efficiency and a highly qualified scientific level. By applied research we expect to find new sources of financing, to avoid, even partially, both internal and external brain drain coming from the low possibilities of public institutions to pay high salaries. I can only mention as an example, the involvement of the Institute of Informatics in applied, profit bringing, projects, available from the evident, very fast increase of demands in this field, submitted from other public sectors, private companies - Albanian and foreign - etc. They find there very fast and highly qualified assistance, introducing the information technologies as a prerequisite for the successful development of their activities. It sounds very pragmatic but the experience, let's say of the Department of Informatics at the University which has practically lost most of the qualified teaching staff - usually going to the private sector often not at all related to their profession- puts the question to have a middle way for protection, within a scientifically oriented environment, of certain socially indispensable potential during a transitional period, satisfying both scientific motivation and economic interest of individuals. In such conditions, the best experts of the Institute cover now also the pressing needs of faculty for teaching. It makes sense to ask why the same strategy is not adopted by the universities themselves? There are some reasons, typically coming from a very wrong approach of the policy of the past to the higher educational system considering universities only as "teaching units" for the education of young generations devoted to the Party. It has created a well-known gap between research and teaching, depriving universities from the real possibilities - in a team sense - to act now as research units and to adapt to new demands which, in a figurative way, as someone once expressed "offer hard currency for soft advice".

The Academy, even in general strategy and not only for the particular illustration I gave here, pays great attention to the collaboration with universities asking from both sides a commitment to participate in joint projects and to provide the university with qualified teaching staff. A special case is the participation and support given by Institutes of R.C. of the Academy to the Tirana University for postgraduate studies with emphasis on PhD thesis preparation.

The intellectual potential of both parts of an Academy can play an important advisory role especially in countries in transition, like Albania: meaning the role of strong personalities who must also have the high ethical reputation as Members of the Assembly and for the R.C. establishing a new approach or style for an efficient science, research and development.

The two parts do not seem to be an obstacle for each other. They will act in synergy for the efficient use of relatively scarce resources available now, compared with great demands needed for the creation of a social and economically advanced society, including the change of mentality in all aspects. This is not an issue to be resolved overnight.

4. In science, research and development, where present and future development of the country are involved and where very skilled and socially adopted human potential is requested, arises the question of the relations between the State and an Academy and if we are talking especially about the financial issue, support provided by state to science.

Hungarian philosophers Feher F. and Heler A. remarked that "in our region, belief on the state paternalism is still very strong"[4]. I am aware that it is also a question of interaction between governmental bodies and the scientific community, but the Academy has to play a key mentality changing role. In particular, in new conditions, the Academy has to prove the efficiency and check the standard of its institutions. On the other hand, in this interaction, as very recently noted by the well-known American economist, Prof.Cuttner, during the 1996 discussions in the U.N. Development Agency, the state cannot renounce from the commitments and responsibilities, in some debated fields, where science is included (from news, USA broadcasted). Combining both remarks, I can only quote the ancient Greek saying "Pan metron ariston", which remains in force as an axiom. I believe that in a democratic state it is achievable. The Academy and its staff have to follow what is prescribed in its functions - professionalism and motivation, responsibility, high ethical attitude to science and scientific truth, to culture and moral values of his nation and other nations as well.

On a regional level, the diversity of types of Academies does not affect a common feature: with their intellectual potential and gained reputation they have come, generally speaking, to be a moving force of social and technological progress. But, unfortunately, we must not forget that, during certain periods, Academies have been under the command of dictatorship. Many times under the honorable name of the Academy, "scientifically" a climate has been created against the spirit of open societies, against the cultivation of real human values of solidarity, against the creation of space for mutual understanding within our region where diversity of cultures could be an advantage and not a source of unreasonable hate. I am convinced that, according to the better climate for an integration to the Community of Democratic States, first of all the European Union, by taking more responsibilities, the Academia can play an important role in this direction.

112

Figuratively speaking, Academies can do much work to invert the meaning of the word "Balkanization" which many times during history was used as a synonym of misunderstandings and the lack of flexibility to find acceptable solutions.

Independently from their specific structure, Academies should - in my opinion - emphasize their efforts for a more effective collaboration on a bilateral basis, for cooperation to resolve many problems of common interest and technically without borders. I used the expression "bilateral collaboration" not as a limitation but as a condition for an effective integration in wider Academic Associations (European, Mediterranean, Black Sea Region etc.). The solution of some of the common problems is becoming vital for our peninsula and as examples I could mention the environment, the biodiversity protection, the energy issue etc. Reinforcing the collaboration and the communication of cultures where they demonstrate the best that they have, we will give to young generations the new conceptions of open societies where they have to respect the dignity of the others as the only way to gain the same for themselves. As institutions of an important part of human creativity, on the eve of the 21st century, the Academies' elected members or scientists working in their institutes, will share and will find proper priorities of their work in the words of Karl Popper "Man has created new worlds - of language, of music, of poetry, of science; and the most important of these is the world of moral demands for equality, for freedom and for helping the weak"[5].

References

1. Theocaris, P.S. (1995) *Academy of Athens and Its Role on Cultural and Technological Development in Greece*, NATO ASI Series, Science and Technology Policy **2**, Kluwer.
2. Kuhn, Th.S. (1973) *The Structure of Scientific Revolutions*, Soc. Edit. **43**.
3. Albanian Human Development Report 1996. UNDP Program, Tirana.
4. Feher. F.Heler,A. (1996) Philosophy of transition, Collection of papers in Albanian.
5. Popper, K. (1996) *The Open Society and Its Enemies*, Rou Fledge and Kegan Paul **1**, 65).

THE SURVIVAL PROBLEMS OF THE ACADEMIES OF SCIENCES OF THE BALKAN COUNTRIES: TO BE OR NOT TO BE ?

S. RADAUTSAN
Academy of Sciences of Moldova
Centre of Semiconductor Materials
5 Academy Str., Kishinev MD 2028, Republic of Moldova

Negative tendencies and contradictions which affected academic science are analyzed - the lack of allocations for investigations, on the one hand, and an enormous number of new problems to be solved, on the other.

New important forms of scientific activities emerging nowadays in the Balkan countries related to different integration processes between academic institutions and industrial, manufacturing or agricultural organizations, are addressed.

The increasing role of a high-level scientific staff in the education and training of young gifted people is emphasized.

Major issues of international scientific cooperation, selection of priority directions for investigations, creation of good conditions for researchers, the role of large scientific centres and the development of science in small countries, are also discussed.

1. Introduction

It is well known that the first academy in the world was that of Athens. A very detailed and eloquent report about the establishment of the first Academy was presented by Prof. P.S. Theocaris two years ago [1].

In 387 B.C., Plato, a prominent philosopher, came up with the idea to create a collective system of education by gathering his disciples in an olive grove "Academus" on the outskirts of Athens. Teaching was effected through the introduction of various tasks which were solved by collective efforts of the Academy's disciples.

Now, 3383 years later, we have also gathered on the outskirts of Athens - this time Saint George Lycabettus Mountain, to discuss a subject of vital importance - "The new role of the Academies of Sciences in the Balkan countries".

Ch. Proukakis and N. Katsaros (eds.),
The New Role of the Academies of Sciences in the Balkan Countries, 113–128.
© *1997 Kluwer Academic Publishers. Printed in the Netherlands.*

Maybe our discussions and recommendations will yield a real contribution to the future development of science and the Athens NATO ARW will be a historic event for the future generations ?

In this communication, I would like to point out several difficulties of academic science at both East-European and national levels, as well as to make some suggestions for solving the problems of promoting and facilitating scientific investigations.

Therefore, the following questions will be exposed:

a) The acceleration of history and the deep crisis of the Balkan countries' scientific organizations.
b) The Academies of Sciences' major problems and new effective forms of scientific activities.
c) The role of academic institutions in scientific training of the young generation.
d) International scientific cooperation and the development of the Balkan countries' Academies.
e) A look into the future. What has to be done ?

2. The Acceleration of History and the Deep Crisis of the Balkan Countries' Scientific Organizations

The changes in our world are speeding up and are bringing nearer the point where the management capacity of political leaders will be surpassed [2]. The acceleration of history at the end of the XX century comes not only from efficient high technologies and faster economic growth, but also from unprecedented world population augmentation and the increasing collisions between expanding human demands and the limit of the Earth's possibilities. For example, Kishinev population grew from 100.000 to 840.000 inhabitants after World War II, the agricultural areas of the Republic of Moldova remaining at the same level. The economic recession that characterizes the transition towards the market economy has led to a considerable reduction of the science budget share in the East European and former Soviet Union (FSU) countries.

Prof. J.E. Aubert [3] analyzed the management of science and research systems in the Balkan countries. He proposed ten main characteristics of the deep crisis and gradual decay of scientific organizations in all Balkan independent states.

We return to these proposals after two years in order to evaluate the level of the crisis in science and to reveal the changes for better or for worse. Let us consider some of these criteria, the most urgent for academic science.

2.1. LACK OF FINANCIAL RESOURCES

In the majority of the investigated countries, the economic situation of science took a turn for the worse. This can be explained by the exhaustion of old reserves as well as by the disastrous state of budgets. The search for foreign support is reduced to humanitarian assistance and grants of international scientific foundations. The drastic shrinkage of the science budget shares is more than two thirds in real terms.

The unexpected rise of fuel materials for meeting energy needs (more than 1600 times) and the discontinuation of foreign economic relations undermined the industrial power and the economic integrity of the FSU states. As a result, the number of investigations was cut by half or even more.

To sum up, the situation at the end of this millennia can be viewed as having the paradoxical similarity with the "boomerang effect": scientific research ensures the acceleration of history which in its turn seriously affects the development of science.

2.2. BRAIN DRAIN

Considerable reduction of R&D personnel takes two forms of brain drain:

a) external-scientists go abroad on a temporary or definitive basis (generally high-level specialists);
b) internal-people leave the scientific sphere to work in the domestic economy.

The experience of the Institute of Applied Physics of the Academy of Sciences of Moldova proves that scientists often are invited to work at foreign universities, while still keeping the contacts with their home laboratories, and also contributing to direct cooperation in scientific research. We can underline here permanent relations of Moldovan professors in physics with the Universities of Konstanz and Darmstadt (Germany), Parma, Rome and Cagliari (Italy), Turku (Finland), Grenoble and Montpelier (France), Ottawa and Montreal (Canada), Athens (Greece), Warsaw and Zabje (Poland).

But there are scientists who have gone abroad permanently, even changing their speciality, emigrating to Israel, Australia, the USA, etc.

The main reasons are financial difficulties, lack of adequate salaries, as well as bad conditions for scientific research, and uncertainty concerning a job in science or education. The average percentage of the external brain drain is about 5-8 %, but it should be stressed again that it refers to the most promising category of researchers, first of all.

The internal brain drain seems to be a more serious concern and it has reached up to 40-50 % of the scientists.

Unfortunately, the situation described by J.E. Aubert [3] can also be observed in the Republic of Moldova - well-trained "brains" work for banks, private enterprises and other important sectors of economy, but a considerable number of Ph.D. holders now work in commercial structures, or as drivers or night-watch men. The most dramatic is the situation of elderly people, pensioners and disabled persons.

There is another category of research workers that was not mentioned be J.E. Aubert [3] - persons who are forced to take an unpaid leave. They sometimes work at their institutes a day (or an hour) per week, their salaries are inadequate and are paid with delay of 3-4 months.

I completely agree with the opinion [3] that "when highly qualified persons accept unqualified jobs, simply to feed themselves and their families, it is dramatic both from an individual and collective viewpoint".

It becomes clear, therefore, that the NATO Scientific Programmes are very important to support international scientific and technological cooperation, to establish personal and professional links, to stimulate investments in long-term projects and introduction of their results into practice [4,5].

Several other criteria of J.E. Aubert will be considered in the text below.

3. The Academies of Sciences' Actual Problems and the New Effective Forms of Scientific Activities

A big advantage of the majority of the FSU Academies of Sciences is the interdisciplinary orientation of their institutes and hence the possibility to solve the most difficult problems which are at the border of different scientific directions.

In the Balkan states, the Academies represent the highest level of scientific authority, the associations of the most distinguished persons - the treasure of each Nation.

The Academy of Sciences of Moldova (ASM), in its 50 Jubilee year (1996), includes 25 institutes, 6 scientific sections, over 1.154 research workers, among them 47 Academicians and 58 Corresponding Members, 150 Doctors Habilitat and 710 Doctors of Sciences (21.01.96).

The ASM has already worked out important measures intended to strengthen the orientation of research towards the demands of the Republic.

Concrete steps are being undertaken to widen relations with departmental science, universities and foreign scientific organizations.

To illustrate this, we can report some priority research subjects performed jointly by the ASM and other institutions:

3.1. THE USE OF RENEWABLE ENERGY RESOURCES

For the Republic of Moldova which has no fossil energy resources, the utilization of renewable energy resources (RER) on a large scale meets 2.5-3% of the annual energy demand. About 45-50% of the rural consumers' demand might be satisfied by RER [8]. The more important sources of the RER could be: solar energy, utilization of urban and agricultural wastes,' wind energy, hydropower of small rivers and geothermal energy. The ASM Institutes of Power Engineering and of Applied Physics together with the respective faculties of the Technical and State Universities of Moldova studied the possibilities to use different RER. For example, a square metre of a photovoltaic cell allows one to obtain an electric power of approximately 45-60 W which can be utilized to supply direct current to household consumers. Solar energy can be converted into thermal energy of hot water and air with the efficiency up to 45-70 %.

The utilization of energy of organic wastes is the next source of energy conservation in the Republic of Moldova.

Agricultural organic wastes constitute a considerable source for biogas through anaerobic fermentation, as well as for the preparation of valuable fertilizers and vitamins of the group B, which are also in insufficient quantities in our country.

The need for international cooperation in urban energy renewal in economies in transition has been recognized as urgent. The creation of a network on integrated urban energy provision in Europe could provide the necessary new impetus for sustainable urban development (K.Brendow [9]).

3.2. EARTHQUAKE PROCESSES IN THE CARPATHIAN ZONE

The studies of the Earth crust structure by geophysical methods of recent and present movements, seismic macrozoning and elaboration of instrumental techniques for seismic and seismotectonic microzoning were performed jointly by two academic institutes: the Institute of Geodynamics "Sabba S. Stefanescu" of the Romanian Academy and the Institute of Geophysics and Geology of the Academy of Sciences of Moldova.

The significance of such investigations is proved by the periodical earthquakes initiated in the deepest crust structures of the Vrantcha mountains (South Carpathian chain), which considerably affected the neighbouring territories of the Balkan Countries, as well as other seismic regions of the globe.

118

It may be important to analyze here the main research directions, the scientific experience and the results obtained by these institutes:

3.2.1. *The Institute of Geodynamics "Sabba S.Stefanescu" of the Romanian Academy (Director Prof. D.Zugravescu, Cor. Memb. of Romanian Academy)*

The Institute studies physical processes that are causally linked to the dynamics of the masses forming the Earth, as well as to the systems Earth-Sun-Moon, and the evolution of the terrestrial deformations in time and space. At present, the Institute of Geodynamics consists of the following units:

- a network of Observatories and Stations grouped in two astrogeodynamic testing grounds;
- a Department for Quick Troubleshooting and Experiments dealing, on the one hand, with the permanent operation of the equipment belonging to the geodynamics observatories, and, on the other hand, with experiments and design of new sensors and instruments;
- a Laboratory for the Calibration and Ageing of the Geodynamics Equipment, situated in Bucharest.

The study of the deformations and of the physical properties of the Earth, the study of the Earth crust, the study of the space-time variations of the geofields and of the way in which these variations influence living creatures - with a special emphasis on the space-time evolution of phenomena that are causally linked to stress cumulations, in geodynamically active areas, which are responsible for earthquake occurrence - have an obvious interdisciplinary character. As a consequence, the Institute incorporated specialists qualified in very different domains and established long-term cooperation with many other research units, with education units, and with production companies with various profiles, from our country and from abroad.

We would like to mention the following areas of international cooperations of the Institute of Geodynamics:

- cooperation aimed at complex geophysical studying and monitoring of phenomena occurring in geodynamically active areas, financed by the European Community, with the Institute of Physics of Earth (Paris), Institute of Theoretical Geodesy (Bonn), Royal Observatory (Brussels), International Centre of Earth Tides (Walferdange);
- cooperation aimed at improving the knowledge of deep structures and modelling of tectonic evolution in geodynamically active areas, financed by the Royal Society (London), Danish Council for Natural Sciences, Deutsche Forschung-Gemainschaft, Romanian Ministry of Science and

Technology, of, respectively, Edinburgh University, Aarhus University, Geophysical Institute (Karlsruhe), and Institute of Geophysics and Geology (Kishinev);

- cooperation aimed at the development of the techniques to reduce the impact of strong earthquakes on cultural heritage buildings, financed by the European Community, with IDROGEO (Trieste), Trieste University and the Institute of Geology and Mineral Explorations (Athens).

3.2.2. *The Institute of Geophysics and Geology of the Academy of Sciences of Moldova (Director - Prof. A.Drumea, Academician ASM) [6,11]*
The main directions of this Institute are:

- the studies of seismic hazards on the territory of the Republic, utilizations of computers and modern devices to accelerate the analysis of seismic events and to establish the earthquake centre parameters [11];
- the investigation of sedimentary rock geology, of the conditions concerning the formation and distribution of construction materials, of minerals and of underground water;
- the geological research aimed at determining the availability of oil and gas deposits in the territory of Moldova.

As regards regional geology, collaborators of the Institute have drawn up the tectonic map of the Republic of Moldova, which is used by different organizations (including the foreign companies) for planning and geological prospecting. Concrete proposal has been elaborated for improving the lands affected by destructive geological processes and for minimizing the danger of soil erosion in the rural areas.

The high-level specialists of this Institute are often invited as experts or for geological research in different countries and continents: Latin America (Venezuela, Mexico, Chili), Oceania (New Zealand), Europe (Balkan Countries, Italy), Asia (Syria, Kuwait), etc. [6,11].

3.3.   BIOLOGICAL PROTECTION OF PLANTS AND MICROBIOLOGY

The problems of high efficiency and ecological safety are of great importance for Moldova's agriculture. The following two academic institutions are involved in these investigations:

3.3.1. *Institute of Biological Protection of Plants (Director-Prof. I.Popushoi, Academician ASM) [6,12]*
The main research directions are:

- the studies of biological methods and comprehensive systems of plant protection, creation of biological means for plant protection and their applications;
- elaboration of procedures for the accumulation of species of insect predators, parasites and photophages for obtaining viral insecticides and microbiological preparations;
- account, identification and synthesis of sexual pheromones and biologically active compounds for the control of pests of agricultural plants and the working out of methods for their use in practice;
- creation of ecological systems of complex plant protection through the use of biological, agrotechnological and genetic methods as well as those of selection;
- elaboration of general recommendations concerning economic and organizational grounds for the application of biological methods in plant protection.

In cooperation with other research centres (Russian Federation, Romania, Ukraine, Germany, USA, etc.), the Institute contributes to research and diffusion of different technologies for obtaining entomophages to control pests.

3.3.2 *Institute of Microbiology (Director - Prof. V.Rudic, Cor. Member ASM) [6,13]*

The main investigations were made in the following research directions:
- microbiological processes in increasing soil fertility;
- microorganisms that can promote plant growth, or produce biologically active substances;
- process of biological nitrogen fixation, pesticides decomposition and improvement of the ecological situation in agriculture [6,13].
- The processes of synthesis of antibiotics, vitamins, pigments, lipids and other biologically active compounds were exposed from the microbiological point of view.

Prof. V. Rudic [13] studies the microbiological compounds obtained by the photobiotechnological methods - microorganisms used to produce protein as a food and fodder, in pharmaceutics, cosmetics, perfume manufacture, etc.

In cooperation with Prof V. Salary, Prof. A. Obuh and other staff members of the Chair of Botany of the State University of Moldova, new promising microalgae strains were obtained to be widely used in medical (high antitumor properties, treatment of cardiac troubles) and zootechnical practice (increase of poultry and pig productivity).

To enlarge the period of the cultivation of microalgae, they have designed an original installation for year-round cultivation of micro-organisms. Installation productivity is two times higher than that now in use. Over two hundred such installations are in use in Moldova, Ukraine and the

Russian Federation providing Spirulina and Dunaliella biomass productions. Of special interest is the biological accumulation of radioactive nuclides (cobalt, caesium, uranium) from various waste waters, using new strains of microalgae as microbiological collectors.

The majority of microbiological investigations have been patented (US and SU Patents).

The total number of biological and agricultural research institutes including the University of Agriculture of Moldova accounts for about 30 entities. Several experimental farms in different districts of the Republic with arable lands, vineyards and orchards are under their supervision.

This branch of Moldovan science is well known and highly appreciated in the world. Implementation of the results constitutes a valuable part of the budget revenues.

3.4. NEW MATERIALS AND HIGH TECHNOLOGIES FOR MODERN ELECTRONICS

The basis of these investigations in solid state physics has been founded by Profs. M.V. Kot and Yu.E. Perlin in the early 1950s at the Kishinev State University. Now in the Republic of Moldova a considerable number of high level specialists work at universities, academic and branch institutions. More than 50 doctors habilitat, about 3.00 Ph.Ds and more than 3000 university degrees' holders in physics, technology, mathematics and other fields have been trained during the last 30 years.

Due to these specialists, electronic and electrotechnical industries have been founded in Moldova. About 15 branch institutes and more than 50 plants deal with the elaboration and application of different electronic materials and devices.

However, the negative processes of unemployment, low salaries, privatization of big enterprises crashed the national economy and the more promising directions of industrial production.

The investigations continued and remarkable results have been obtained in growing amorphous semiconductors for optoelectronics [5], in obtaining thin films and fibers of ternary and multinary semiconductors and high-T_c superconductors, in studying their structural, mechanical, electrical, optical and magnetic properties for practical applications [4]. The results have been reported at international conferences [14-16] and published in specialized books [17-18].

Prof. S. Rowland, Molina and P. Crutzen, specialists in atmospheric chemistry, were awarded the Nobel Prize for their discovery in the early seventies of chlorofluorocarbones that could deplete the thin layer of stratospheric ozone that protects life on Earth from harmful ultraviolet radiation. All humanity is indebted to them for forecasting the change in the ozone layer just in time to save it.

We studied the semiconductor ternary materials in the system zinc-indium-sulphur ($ZnIn_2S_4$, $Zn_3In_2S_6$, etc.) with a large energy gap and on its base we constructed the ultraviolet dosemeter for uses on land and in space [16]. The same semiconductor devices are promising for applications in medicine, agriculture, industry, etc.

At the NATO ARW in Kishinev [4,5] at Yokohama and Stuttgart [15,16], more than 30 reports were presented concerning the results of investigations in photonics, laser technique, fibre telecommunications, new technologies in material research science. A considerable number of the reports were accomplished with co-authors from different foreign countries.

One eloquent example here is the following one. For the first time, an original magnetic material $ZnMn_2As_2$ was grown in Kishinev (Moldova) and was reported. It attracted the attention of research workers from Germany, Belgium and France. As a result, an international team was set up to study this ternary compound [19].

Frontiers in High Magnetic Fields may induce Frontiers in Materials Engineering and vice versa, especially for materials combining semiconductor and magnetic properties in semimagnetic semiconductors. $ZnMn_2As_2$ is a new material of the class $(Zn_{1-x}Mn_x)_3As_2$. For x=2/3, however, the usual statistical substitution of the Zn-ion by Mn^{2+} condenses into an ordered layer structure. The different magnetic phases are determined by neutron scattering, compared with low temperature magnetization data and detected directly in the DC- magnetoresistance of the low mobility holes, whose conductivity properties are essentially determined by the internal spin polarization of the material.

Another example is from a different sphere. An original direction of investigation was initiated by studying the archaeological objects of the Dacian civilization by the methods used in microelectronics. Exciting results were obtained, some of them being reported at the two NATO ARWs [4] and the present one.

3.5. SOME NEW EFFECTIVE FORMS OF SCIENTIFIC ACTIVITIES

The legislative basis for research and developments (R and D) is not completely finished yet, due to a rather short period of Moldavian independence.

The Parliament of the RM adopted some laws which concern science problems. They are the following:

- on education
- on author's certificates and patents
- on standards

- on foreign investments and support of innovators
- on trademarks and signs
- on excises
- on fiscal policy and land taxes
- on road taxes
- on wine and viticulture, etc.

But it is very difficult to set up legal frameworks which are basic to a market economy [3].

For the evaluation of the science systems' level for the selection of the best institutions and scientists and in order to support them, the President of the Republic of Moldova formed the Consultative Committee for Science and Sustainable Human Development in which well known scientists of Moldova have been included.

The International Council of Scientific Unions (ICSU) has provided seminars to diffuse the practice of peer reviews, which is necessary for properly administering grant funding or selecting priority areas. The ASM is a member of ICSU [24]. It also participates in various European and world forums.

4. The Role of the Academies of Sciences in Training Future Generations

From our point of view, one of the most important missions of the academies, from Plato's [1] till the modern ones, [6] is to educate the future generation of the researchers.

The majority of the FSU academies have had post-graduate departments and networks of specialized councils for conferring scientific degrees. Moreover, a system of the scientists' promotion and post-graduate study was organized in the famous academic institutions.

At the Academy of Sciences of Moldova, 17 specialized councils were organized which were attached to 14 institutions. The total number of post-graduate students per year varied between 90-110 persons, besides the fact that 10-15 persons had the possibility to prepare their Ph.D. theses abroad. During a long period, academic institutions helped the universities and other branch scientific organizations offering high-level specialists - lecturers in respective fields. Nowadays, about 1/3 of young doctors after the defense of their theses begin to work for foreign firms on contracts (as football or hockey stars). Maybe it would be correct to pay a part of the contract's sum to the institutes or universities where the specialist has been trained ? (The effort and time for training a scientist is much longer and more difficult than training a sportsman.) It should be mentioned that the academic conditions for talented young specialists are very effective.

124

An eloquent example can be given here. Sergiu Vacaru (born in 1958, Briceni distr., Moldova) is a Senior Researcher, Doctor in Physics and Mathematics. His fields of interest and scientific activity are theoretical and mathematical physics and geometry:

- gravitational waves and nonlinear optics;
- twistor and spinors, gauge theories and gravity;
- spin glass state and high temperature superconductivity physics;
- generalized Lagrange and Finsler geometry supergeometry;
- stochastic processes in curved spaces;
- mathematical modelling in economics
- Publications (1983-96): [22]

More than 90 scientific papers and contributions at international conferences (18), Monographs prepared for West countries (2) (Springer Verlag, Academic Press)
Education and Positions:
1976-82 Tomsc Politechnic University, Russia
1982-84 Joint Inst. Nuclear Researches, Dubna, Russia
1984-87 M.Lomonosov State University, Moscow, Russia
1987-96 Scientific worker, Acad. of Sciences of Moldova
1988-89 T.Shevchemco State Univ. Kiev, Ukraine
1992-94 "Al.I.Cuza" Univ., Iasi,Romania Ph.D. theses on theoretical physics.

So, it all depends on the person. Still, another important point here is that there should be incentives for future Nobel Laureates, and these incentives should be tailored. They could be a better job or promotion in well-known world scientific centres, or maybe some others. All these ideas are open for discussion, and any new one will be gratefully appreciated.

5. International Scientific Cooperation for the Development of the Balkan Academies

A large scientific cooperation of the Balkan Countries is organized in the framework of the Black Sea Economic Committee relations as well as under bilateral programmes [23, 24]. A long-term and fruitful cooperation was carried out including 168 contracts (1995), in different fields of investigations. The distribution of financial funds for the research directions was:

1.	Applied Physics	14.6 %
2.	Electrotechnics, Electronics	24.7 %
3.	New Technologies and Materials	17.7 %
4.	Agriculture	16.2 %
5.	Medicine	11 %
6.	Biology	6.4 %
7.	Ecology	3 %
8.	Other subprogrammes	3 %

The part of the academic institutions constituted about 43.4 %. Most active were the Institute of Applied Physics, the Institute of Chemistry, the Institute of Genetics, the Institute of Geophysics and Geology, the Technical University of Moldova, the National Institute of Wine and Viticulture, etc. [23].

Presently, Moldova is participating in the realization of some projects under the UNDP, UNIDO, TACIS, INTAS and other international grants. At the last Forum of the Black Sea Economic Cooperation (Bucharest, April 1996) [24], special attention was paid to the industrial impact on the environment. It was stressed that a healthy world is possible in the future only with clean industry.

On November 4-14, 1996, in Brussels (Belgium) the World Salon of Inventions, Investigations and Transfer of Technologies was held; 45 countries presented 1200 works. The delegation from the Republic of Moldova (18 persons) presented 26 plane-tables with 54 inventions and patents. They were highly appreciated:

- 10 received gold medals,
- 5 received silver medals,
- 4 received bronze medals.

At the special reception in the Mayor's Office, Professors V. Rudic and I. Gulea were decorated with the Cross of the Kingdom of Belgium for distinguished merits at 3 Salons consecutively. At that reception Academician I. Bostan was decorated with the Royal Medal for distinguished merits.

At that Salon, various contracts of scientific cooperation were under discussion: in the field of precision planetary transmission, with AIRBUS, ARIAN, with the NATO Scientific Council.

6. A Look into the Future. What Has to be Done ?

In the not very distant 1967, I had an opportunity to visit six universities and some research centres of the USA. At the Washington Institute of Problems of America, Prof. Tshern, a specialist in Soviet problems, discussing the future, forecasted disaster for the countries with a very militarized economy. He was very prophetic. Today's situation in the Balkan countries is the realization of his prognosis.

The year 1968 marked the end of the long post-war period of rapid economic growth in the developed countries, but it was the year when social unrest and public awareness of the problems of the environment began to emerge.

The Club of Rome arose in 1968 from these considerations. It comprised one hundred members from fifty three countries that united in a common concern for the future of humanity. The Club's thinking has been governed by three related conceptual patterns [25,26]:

- to adopt a global approach to the vast and complex problems of the world, in which the independence of nations within a single planetary system is constantly growing;
- to focus on issues, policies and options in a longer-term perspective than is possible for governments, which respond to the immediate concerns of an insufficiently informed constituency;
- to seek a deeper understanding of the interactions within the range of contemporary problems - political, economic, social, cultural, psychological, technological and environmental - for which the Club of Rome adopted the term 'the World Problematique'. The Club lived up to its role.

Today, the situation in scientific research and, first of all, in fundamental investigations is aggravated.

Therefore, I propose to organize immediately a Club of Athens to save science in the East European and Balkan countries (and not only) and its heart - the Academies of Sciences.

The primordial problems to be solved by the Club of Athens may be the following:

- to collect the intellectual World Treasure "the grey golden cells"- the data about eminent scientists of different specialities and countries, and to organize the international protection of these persons with the help of UNESCO, UNIDO, etc;
- to create a fund for the assistance to young scientists;
- to organize together with the NATO Department of Science biannual specialized workshops with the participation of promising young scientists, as guests, or maybe one or two of them could even be key speakers.

A very important task of the Club will be to highten the prestige of science in the Balkan countries. It will be necessary to use the media, to contact scientific popularization organizations (e.g. "Znanie") and these groups of persons who are VIPs in political life [3]. It could take into account not only scientific achievements of research workers but also the contributions they are making in the economic and social renewal, in the intellectual wealth, environmental protection, education systems, etc.

To summarize, the answer to the question put forth in the title of this report is only one and optimistic - for the academic science in the Balkan Countries - to be !

References

1. Theocaris, P. (1995) The Academy of Athens and its role on the cultural and technological development in Greece, in G. Parissakis and N. Katsaros (eds.) *Science Policy and Research Management in the Balkan Countries*, Kluwer Acad. Publ., Dordrecht, pp. 1-8.

2. Brown, L.R. (1996) Acceleration of history, in L. Starke (ed.), *State of the World, 1996, A Worldwatch Institute Report on Progress toward a Sustainable Society*, W.W.Norton and Comp., N.Y., pp. 1-27.

3. Aubert, J.E. (1995) Science in the Balkan countries; key policy issues, in. G. Parissakis and N. Katsaros (eds.), *Science Policy and Research Management in the Balkan Countries*, Kluwer Acad. Publ., Dordrecht, pp. 67-78.

4. *Scientific and Technological Achievements Related to the Development of European Cities* (1996), *NATO ARW Abstracts*, G. Parissakis and S. Radautsan (eds.) Kishinev.

5. *Physics and Applications of Non-Crystalline Semiconductors in Optoelectronics* (1996), *NATO ARW Abstracts*, M. Bertolotti, A. Andriesh (eds.), Kishinev.

6. *Academy of Sciences of Moldova* (1993), A. Andriesh (ed.) Stiintsa, Kishinev.

7. *Academy of Sciences of Moldova - 50* (1996), Informative Guide, H. Corbu (ed.), Stiinta, Kishinev.

8. Andriesh, A., Kiorsac, M., Simashkevich, A. (1996) The opportunities of using of renewable energy resources in the Republic of Moldova, *"Eurosun-96" Intern. Conf. Abstr.*, Friburg, pp.1-6.

9. Brendow, K. (1996) Energy efficient management of cities: a review of approaches worldwide including the economies in transition, in S. Radautsan and G. Parissakis (eds.) *Scientific and Technological Achievements Related to the Development of European Cities*, Kluwer Academic Publishers, Dordrecht, pp.63-86

10. *Abstracts of the Geophysics Intern. Symp.* (1995), D. Zugravescu (ed,) Inst. of Geodynamics "S.Stefanescu", Romanian Acad., Bucharest.

11. Drumea, A.V., Shehalin, N.V., Grafov, S.S. (1990) Carpathian Earthquake of August 31, 1986. Kishinev (in Russian).

12. Popushoi, I.S., Voloshtchuk , L.T. (1994), *Biotechnology: Realizations and Perspectives in the Republic of Moldova. Bul. ASM*, N 5, Kishinev.

128

13. Rudic, V. (1995) Achievements and prospects of photobiotechnology in Moldova, in G. Parissakis and N. Katsaros (eds.) *Science Policy and Research Management in the Balkan Countries*, Kluwer Acad. Publ., Dordrecht, pp. 131-139.

14. *Ternary and Multinary Compounds. Proc.ICTMC-8* (1990) S. Radautsan, C. Schwab (eds.), Stiinta, Kishinev.

15. Radautsan, S., Tighinyanu, I. (1993) Defect engineering in II-III-YI and related compounds, *Jap. Journ. Appl. Phys.*, **32**, Suppl. **32-3**, pp.5-4.

16. Radautsan, S. (1995) Achievements in solid state electronics in the Moldova Republic, in G. Parissakis and N. Katsaros (eds.) *Science Policy and Research Management in the Balkan Countries*, Kluwer Acad. Publ., Dordrecht, pp. 119-130.

17. Ghitsu, D., Kantser. V., Popovich, N. (1986) *Ternary Semiconductors with Low Energy Gap*, Stiinta, Kishinev (in Russian).

18. Andriesh, A., Bivol, V., Buzdugan, A., et al. *Semiconductor Glasses in the Photoelectrical Systems for Information Recording*, Stiinta, Kishinev.

19. Ortenberg M. von et al. (1994) The application of high magnetic fields in solid state physics, *Physica B*, **201**, pp. 57-62.

20. Urekian, S. (1996) An important aspect of municipal management: the sociopolis strategy, in S. Radautsan and G. Parissakis (eds.) *Scientific and Technological Achievements Related to the Development of European Cities*, Kluwer Academic Publishers, Dordrecht, pp. 1-12.

21. *City Centre for Housing Assistance* (1996), Moscow Govern., Engineering Provision Dept., Moscow

22. Vacaru, S., Ostaf, S. (1996) Nearly autoparallel maps of Lagrange and Finsler space, in P. Antonelli and R. Miron (eds.), *Lagrange and Finsler Geometry*, Kluwer Acad. Publ., Dordrecht.

23. *Scientific and Technological Cooperation between Romania and the Republic of Moldova* (1996) F. Tanasescu and S. Radautsan (eds.), Ministry of Research and Technology, Bucharest.

24. Andriesh, A. (1996) Scientific achievements related to the Sustainable Development of cities, in S.Radautsan and G.Parissakis (eds.) *Scientific and Technological Achievements Related to the Development of European Cities*, Kluwer Academic Publishers, Dordrecht, pp. 13-21.

25. Meadows, D.L. et al. (1972) *The Limit of Growth. A Report for the Club of Rome's Project on the Predicament of Mankind*, Univers Books, New York.

26. King, A., Schneider, B. (1991) *The First Global Revolution. A Report by the Council of the Club of Rome*, Simon and Schuster Ed. London, Sydney, New York.

SCIENCE AND INTERNATIONAL COMMUNICATION

The role of Academies of Sciences: Building Bridges.

P.J.D. DRENTH
Royal Netherlands Academy of Arts and Sciences

The first Academy, founded by Plato in the fifth century before Christ, was situated in the grove near Athens outside the centre of public life. It was not the wish for isolation but rather the desire of independent individuals to engage in critical reflection on both the prevailing philosophical theorems and the political arguments that made that place most appropriate. When the Emperor Justinian almost a millennium later decided to close this Academy, since he believed that the views developed and disseminated there were damaging to the state, he did not realize that the very value and contribution of an Academy does spring from its independent position and its freedom to criticize. Later, in the Renaissance, it was this same spirit of independence and intellectual freedom that caused the revival of the idea of an Academy. An academy is a place where scholars meet, exchange ideas and reflect in an environment and spirit of absolute intellectual freedom and independence.

Of course, these intellectual activities produce ideas, thoughts, critical reflections and recommendations, which governments, institutes of learning and research or funding agencies may want to use within the realm of their own objectives.

An Academy of Arts and Sciences, therefore, is a meeting place for scholarly and scientific reflection, generating ideas and intellectual products which may prove useful for the society at large. Here we have in a nutshell the very mission of an Academy, however different this has been and is operationalized by the various national Academies in Europe.

Before we focus our attention on the subject of our debate I would like to discuss a terminological question: The theme of this conference apostrophizes science: the role of the Academies of Sciences in the Balkan countries. But the English word "sciences" has a more restricted annotation than for instance the Dutch term "wetenschappen", the German "Wissenschaften" and the Scandinavian "videnskaber". The English "science" refers to the natural sciences primarily. In many countries this has led to separate Academies for natural sciences and the humanities. In the Netherlands we are fortunate to have

Ch. Proukakis and N. Katsaros (eds.),
The New Role of the Academies of Sciences in the Balkan Countries, 129–137.

just one "Akademie van Wetenschappen". I say fortunate, since it is my conviction that what unifies sciences and humanities is more significant than what divides them.

In this line it is also my opinion that the debate should be extended to the whole field of learning and scholarship, including the humanities and social sciences. In the position which I will take in this contribution I will, therefore, refer to this wider concept of "scientia". And particularly with an eye on the theme of my paper: how can we promote communication and understanding between peoples, the inclusion of humanities and social sciences is, to my opinion, indispensable.

In the following we will analyze the three functions of an Academy of Sciences, and demonstrate that all three may help to build bridges and to improve communications and to reduce political or even military tensions between nations or regions.

Forum function

In the first place, Academies have the important and central forum or meeting function. Gatherings of the general assembly, divisions or sections take place on a regular basis. In addition, special meetings, conferences, colloquia and workshops are organized by the academies or under their auspices. Also the international contacts, reciprocal visits of scholars, special lectures, exchange of periodicals and other information, membership of international organizations, including ICSU, ESF, UAI, ALLEA and others, emphasize the international nature of the meeting function.

It will be obvious that in this capacity Academies will have an important bridging function. All through history, science has been international. True scientific discourses never have bothered about national boundaries. Even in the darkest times during the repressive Stalinistic regime scientists from Russia communicated with the scientists from the West. In fact the contacts between academies were often the only avenue for these communications.

Of course, there are, sometimes even sharp, differences of opinions between (groups of) scientists. In fact, controversies, debates or even clashes of opinions are fruitful and stimulating. They help to keep the discussion going and the discourse alive.

But two important observations have to be made in this respect: In the first place, these differences in opinion seldomly coincide with divisions between nations. Secondly, although the debates are sometimes heated and fierce, these differences are basically agreeable to reason, and solvable - or at least discussable - in rational logical terms. Scientific differences of opinion and controversies can only be solved through reason and arguments, and never by means of power, force or hostilities.

In such a dialogue it is the common search for the truth, the attempt to apprehend each other's arguments, and the joint effort to analyze and to comprehend the various issues and arguments which unite rather than divide, and which lead to a common understanding.

Research function

Secondly, Academies of Sciences have an important responsibility in stimulating and promoting scientific research. This can take place either directly by sponsoring or executing research programmes, or indirectly by playing a major role in the evaluation of research by producing research foresights in different disciplines or through the appraisal of individuals or institutes in order to award prizes, scholarships or distinctions.

As far as taking responsibility for own research programmes is concerned, it is clear that Academies differ extensively with respect to the question whether they should incorporate or create *research institutes*, or whether they should restrict themselves to the intellectual discourse and reflection upon the research work of their own members and scientist outside the Academy. In the latter case the Academy will have a rather modest administrative staff, just to support its meeting and advisory function (which we will discuss later) and in most cases a library. This is the type of arrangement to be found in many Western European countries. The basic idea behind this approach is that scientific education and scientific research should not be separated, since they need and mutually support each other. Consequently, (fundamental) scientific research should in principle be carried out within universities.

Until recently, in Eastern Europe, the Academies often accommodated very large numbers of research institutes. In fact, the bulk of fundamental and applied research in those countries was to be found within the Academy institutes and not within the university system.

There are some modalities in between. As an example, I can present the case of my own Academy, where we have 18 institutes ranging from very small (4 to 5 persons) to large (over 200 collaborators). The criteria for an institute to be incorporated in or to be created by the Academy include:

- high scientific level;
- concerned with problems or issues of national interest;
- studying problems areas that are not or cannot be studied by a particular university or another national research institute.

Possible conflicts between the research/administrative responsibilities of an Academy and its advisory role should be acknowledged and dealt with; in our case by special organizational and management arrangements.

Whether an Academy is doing its own research or restricts its involvement to evaluation, stimulating and sponsoring research of others, in any case it deals with an activity that presupposes collaboration and contacts, the exchange of knowledge, expertise and research results and attempts to apprehend each other's work. And again - and this is an important consideration given the theme of this presentation - these contacts, collaboration and exchange have to cross national borders. Science is an international phenomenon; this has always been the case throughout history, but has become particularly conspicuous in modern times with its fast and sophisticated communications and information systems. The term "national science" is almost a *contradictio in terminis*. Even for national research programmes it is a "must" to have international connections and to participate in international collaborative networks.

The following arguments can be brought forward to substantiate this position:

- Mondial responsibility. Some of the (major) international research programmes and projects can only be initiated and supported if sufficient international partners are involved. It is a moral obligation for countries that are capable to contribute and to participate to do so.
- To "keep in touch". It is essential for scientists and researchers in any country to keep (also personal) contact with developments elsewhere. Communication and cross-fertilization is essential for the scientists' own motivation and for the education and training of younger scientists. Moreover, participation in an international research effort is often a requisite prelude to further national or local research.
- Many research issues have a supra-national scope, and less and less problems can be studied fully from a purely national perspective. Research areas such as environment, health (transferable diseases, aids), energy, transport, tourism and trade, banking and finance, and migration come to one's mind easily.
- With respect to certain international questions a particular country may have a specific interest, because of the special national needs. These questions may be given priority by a country on strategic grounds. As far as the Netherlands is concerned special emphasis is given to transport and trade, telecommunication and other forms or communication (linguistics), high-tech electronics, health research and environmental research.
- Finally, there may be some research areas where a country may be able to provide a unique contribution because of its specific expertise and experience in these fields. Again with respect to the Netherlands, this may be relevant for research in the field of civil engineering (water control), agriculture and fishery, fresh and salt water interface, astrophysics, microbiology and biotechnology.

Thus, through the requirement to consult and to collaborate with each other in *international* research programmes, and to have to exchange knowledge and ideas and to take cognizance of results and achievements of colleagues throughout the world in more *national* or *regional* research activities bridges across borders are built and enforced. And it is beyond a doubt that Academies, in their fostering this research, are salutary in this process.

Advisory function
The third, and for the purpose of our discussion probably the most instrumental function of an Academy of Sciences pertains to its advisory capacity.

This advisory role addresses questions related to science and science policy, ranging from specific arguments to general (e.g. ethical) issues, from fundamental scientific controversies to problems of technological, medical, legal or political applications.

It is particularly the latter that has to be elaborated in the context of my paper. At stake is the issue of relevance of research; the question of how an Academy can steer science and science policy or applications of scientific findings towards beneficial societal objectives such as peace, understanding, justice and well being of the people. How and to what extent can research, be it pure or applied, be qualified as relevant?

Much of the recent political debate on the appropriateness of scientific research and, more importantly, the justification for its funding is rooted in a rather narrow definition of relevance as the contribution to economic growth and technological development. Research utility is often seen as amounting to its capacity to stimulate the growth of the gross national product. This point of view is, of course, far remote from the one propagated by the Neo-Marxist ideology in the 1970s, in which relevance was equal to the extent to which science (and particularly social science) contributed to the emancipation of the lower classes, to the redistribution of power in organizations and to the general ideals of equality and a free and democratic society. At the same time, it distinguishes itself from relevance defined in simple instrumental terms: the extent to which scientific research can help to build strong dikes and bridges, to develop powerful drugs against dreadful illnesses, to construct effective learning devices, or to assist the marketing department of a food chain to sell its beans or dairy products better. Clearly, relevance and usefulness may have many different faces, may serve many different objectives, and may benefit many different utilizers.

I would like to distinguish three dimensions in the "relevance space".

1. First of all, I want to refer to a conceptualization of relevance which goes beyond the (often short-term) economic value and practical applicability: *intrinsic relevance*. Raising questions on the origin of the earth, on the nature of matter, the sense and meaning of life, the essence of communication between creatures, the shaping of values and norms, and the like, is a fundamental and unique characteristic of the human species and a motor for its development. Any developed and civilized society should acknowledge the importance of attending to such existential questions, and they should take pride in promoting (and, I should add, in financially

supporting), posing and answering them. Research, be it in the natural sciences, in the humanities, or in the social sciences, leads to an augmentation of the body of knowledge, an intrinsically valuable and precious quality of civilization itself.

There is one other thing: science, both fundamental and applied, also has an important *educational* function. Wherever good research is being carried out, one also finds young researchers being trained, either formally or informally, for their own academic or research career. Given the importance of the preparation of the next generation of high-level and well-trained scientists and scholars, this aspect of the intrinsic relevance of scientific research also deserves special appreciation.

This educational function has an even broader dimension; intolerance, extremism, xenophobia and intergroup conflicts are often a product of ignorance. Therefore, the educational function also pertains to the broader public, and the (scientific) education of the general public can be considered to be an important instrument to develop and to strengthen the intellectual defensibility and democratic foundation of a society.

In this respect it is particularly the social and behavioral sciences which deserve an important position. They not only deal with essential dimensions of human and social life, which we need to know in order to understand the behaviour, the interactions and the social structures of modern man. But we also have to train future generations of scientists to further explore and analyze these essential human and social dimensions, and to educate next generations of the general public, in order to strengthen future democracies.

2. Secondly, there is, of course, *instrumental relevance* the immediate or indirect application of research through the transformation of its findings into practical and useful tools and instruments. The construction of tests or learning devices, drugs for therapeutic or prophylactic use, aids for perception- or motor-handicapped people, and a great many other applications can be listed to illustrate this point for my own discipline psychology.

Somewhat less directly technological, but still falling within this functional definition of relevance, is the contribution which scientific research can make to the development or creation of insights or new knowledge which lead to important breakthroughs in preventive or therapeutic approaches or in developmental or intervention practices.

It should be emphasised here that, while this instrumental relevance is often a product of what is called applied research, this is certainly not always the case. In fact, the research which leads to this functional use is frequently not aimed at practical applications at all. One may think of the time lag between the development of the Radon theory 60 years before its application in computer topography, the purely theoretically inspired development of polymer chemistry some 40 years before its application in plastics manufacture, and the development of the telegraph by Marconi 30 years after Maxwell's fundamental work on the transmission of electromagnetic waves.

3. A third form of relevance can be called *contributive relevance*: Scientific information and knowledge can contribute to better judgement, to more appropriate decision-making and to more suitable interventions in a variety of contexts.

The actual process of influencing or convincing decision-makers can take various forms: In the case of economic decisions, it can be purely *cognitive* and rational. In other cases, scientific research can create a positive or negative *attitude* towards a possible decision option, for instance by showing that there is a high probability of dreadful outcomes resulting from an intended decision. Relationships between smoking and the occurrence of cancer or heart diseases, or between watching violent TV programmes and criminal behaviour, are cases in point. Finally, research results can be used in a *debate* on political decisions or preferences, to convince an opponent or to weaken his position. For instance, research evidence on the negative relationship between family income and family size in developing countries, or on the negative relationship between education and population growth, may be powerful arguments in the justification of aid and development programmes in these countries.

It will be clear that for a proper understanding and appreciation of a variety of societal or political problems, including international tensions and conflicts, a solid, critical and in depth scientific analysis is preconditional. Additionally the Academies can utilize this scientific knowledge in trying to create better and more responsible political attitudes and to promote a more accountable political decision making. It is my conviction that for many of these problems an Academy is a suitable candidate for providing such analyses and advices.

This advisory capacity of an Academy can be defended on three grounds. In the first place the availability of the abundant scientific knowledge and experience within its walls. The country's top scholars and scientists form a large reservoir of expertise which can be utilized for advice and consultancy. Secondly the impartiality of the Academy members. No political, economic, regional or professional interest group can nourish the hope of being especially favoured in the Academy advice. Thirdly, the exclusive scientific orientation of the Academy and their advisory work. It is just finding out the truth that is of interest to them.

In the Netherlands, the Royal Academy of Arts and Sciences has appointed a broad permanent "Advisory Committee on Science and Ethics", which exercises a "forum"-function, among others through the following tasks:

a) Organization of yearly symposia on "Science and Ethics", resulting in publications or reports for a wide public.
b) Contribution to the national platform-discussions which the Minister of Education and Science has initiated, and which focuses on questions of the impact and limits of science and technology.
c) Inclusion of the socio-ethical dimension in its criteria for the evaluation of research programmes within its own institutes and with respect to research carried out by universities or institutes outside its own realm.
d) Contribution to and participation in existing national advisory committees on science and ethics, such as the committees on animal experimentation, biotechnology and genetic engineering, technology assessment and the like.

A caveat should be formulated here. The Academies of Sciences have always to realize their specific position and responsibilities. Scientists should not jump further than the length of their jumping-pole. They could offer proper and careful analyses of the problems at hand, including the complexities, multicausalities and non-linear relationships. They could provide a definition of alternative options, they could give empirically based estimates of various "if-then" relationships, but they should not move actively into the political field themselves. Their trade is science and they should not become another political pressure group or take over the responsibilities of the actual responsible decision makers.

In other words, there is a limit to the moral obligations of scientists as scientists. They should not take more, and in fact less, responsibility than politicians, employers, therapists and doctors. They have to examine the social implications and consequences of their research, they have a responsibility in educating people to be aware of these consequences, and in developing proper guidelines and safeguards, but they should not be more socially accountable than the actual responsible decision makers. If it not up to the physicist to decide whether or not an atomic bomb is to be built; it is not up to he epidemiologist to decide whether carriers of infection viruses should be kept in quarantine; it is not up to the behaviour scientist to decide whether incurable and deeply depressed patients should be allowed a eugenic termination of their life. There is a danger in asking too much responsibility of scientists. It would shift too much of the power from responsible decision makers to people that are neither trained nor (as such) competent to exert it.

Conclusion

By such a scientific analysis of political or societal issues, such a participation in the public debate and such an advisory responsibility the Academy of Arts and Science can contribute significantly to rational approaches in finding solutions and to a political decision making that is agreeable to reason.

Through the three functions of Academies as described in the foregoing they can build intellectual and spiritual bridges between edges, even if these may seem to be separated by rather profound abysses.

Recognizing the important and distinctive potential of science and scholarship to contribute to a better and peaceful future of mankind the Academies should adhere to the two general principles, phrased in the 1996 Genoa Declaration on Science and Society as follows:

- "Respect for the diversity of cultures within societies and promotion of science as a distinctive and important contributor to bridging such diverse cultures and promoting peaceful coexistence in accord with the principles of freedom, autonomy and rationality.
- Mutual cooperation, reflecting the recognition that the production and utilization of scientific and technological knowledge are decisive for the future welfare of humanity and that science, with its universality, is uniquely positioned to serve as a laboratory in which mankind can work together to achieve a better future in accord with the principles of responsibility, solidarity and respect for the rights of individuals and nations."

ACTIVITIES OF THE CANADIAN ACADEMY OF SCIENCE AND OF RECENT INNOVATIVE CENTRES IN CANADA

B.P. STOICHEFF
The Royal Society of Canada
225 Metcalfe Str., Suite 308, Ottawa, ON, Canada, K2P 1P9

1. Introduction

Academies of Science, from their earliest beginnings, have played major roles in each of their countries and for humankind in general. Each has had specific goals, and from time to time, each has enjoyed the respect and support of its citizens and governments. Nevertheless, in response to present worldwide structural changes, it is not surprising to find that Academies in many countries are re-examining their roles and responsibilities.

Another current trend is the formation of groups of academies at the national, regional and global levels[1]. For example, the four academies in the UK (the British Academy, the Conference of Medical Royal Colleges, the Royal Academy of Engineering, and the Royal Society) established the National Academies Policy Advisory Group (NAPAG) in 1992 to facilitate collaboration on multidisciplinary policy issues. Similarly, groups of academies have been formed in Germany, in Switzerland, and in Finland, to promote mutual cooperation and to represent them internationally. At the regional level, the Federation of Asian Academies and Societies (FASAS) was established in 1983 and now has 12 members; in Europe, ALLEA (All European Academies) was initiated in 1990 and has a membership of 53 academies. At the global level, the Council of Academies of Engineering and Technological Sciences was set up in 1978 and has grown to 13 members. More recently, the InterAcademy Panel (IAP) was formed with 57 academies as members, to focus on international issues, such as population and development.

Thus, the present workshop is an auspicious occasion when the Balkan Academies not only rise to the challenge of new roles, but may also embrace the benefits of developing worthy collaborative endeavours in this part of the world. For my part, I propose to summarize the activities of the Canadian Academy of Science, and of several new institutions developed in Canada in the past decade, with the hope that these examples may strike a common

Ch. Proukakis and N. Katsaros (eds.),
The New Role of the Academies of Sciences in the Balkan Countries, 139–146.
© 1997 *Kluwer Academic Publishers. Printed in the Netherlands.*

chord with the attendees and suggest possible new approaches to the problems we face in many of our countries.

2. The Academy of Science of the Royal Society of Canada [2]

The Academy of Science is one of three academies of The Royal Society of Canada, the others being, the Academy of Humanities and Social Sciences, and Académie des lettres et des sciences humaines. These bring together Canadian scholars in a wide range of disciplines. Founded in 1882 and modelled after The Royal Society of London and the Institut de France, The Royal Society of Canada (R.S.C.) is an independent, self-governing organization supported by its members, governments, and public and private sources. Election to the Society is an honor, and the, approximately, 1400 Fellows have been elected in recognition of distinguished contributions to scholarship.

The objective of The Royal Society of Canada is the promotion of learning and research in the arts and sciences. It recognizes distinguished accomplishments and provides timely information and advice to governments and the public. As an organization of eminent scholars and scientists, it forges vigorous partnerships to seek solutions to urgent and complex challenges of the day.

The Academy of Science is the largest of the three academies, and currently comprises a Fellowship of 800 members organized into four divisions:

- Applied Science and Engineering
- Earth, Ocean and Atmospheric Sciences
- Life Sciences
- Mathematical and Physical Sciences

The main objectives of the Academy of Science are:

- to promote learning and research in science and engineering
- to recognize and honor excellence in scientific research
- to foster a greater appreciation for science in Canada
- to provide advice on national and international scientific issues
- to represent and support the scientific community
- to inform the public on important scientific issues of the day

These objectives are largely met through a variety of activities, often taken together with the other academies; for example,

- regional meetings
- symposia

- awards and medals
- contract studies
- public awareness of science
- lecture exchange programs

Major studies being carried out by the Academy of Science, some under contract and some in collaboration with the two other academies include:

- The Canadian Global Change Program, now in its second decade. It bridges the science and policy-making communities in Canada in the area of global environmental change, and has as its mission the promotion of informed action in support of sustainable development at the public, corporate, and individual levels through sound advice on global change. Over the past several years, the program has sponsored many public events and symposia dealing with global change issues, and has published a newsletter and booklet on these topics.

- Review of the Nuclear Fuel Waste Management and Disposal Concept This joint committee with the Canadian Academy of Engineering has held public hearings and completed a report on their recommendations to the government.

- Public Awareness of Science Program has a mandate to promote a science, innovation, and entrepreneurial culture in Canada, including the social as well as natural sciences. The present focus is on network developments with links to any organization interested in public awareness of science. Lecture series have been organized for the National Science and Technology week.

Other studies and activities of the Society, carried out by volunteers, include the work of specific committees on:

- Freedom of Scholarship and Science
- Women in Scholarship
- International Decade for Natural Disaster Reduction
- International Relations

Lecture exchange programs have been established with several world Academies, leading to visits of distinguished scholars to each other's country in alternate years:

- The Royal Society of London and RSC - natural and life sciences
- The British Academy and RSC - humanities and social sciences
- Académie des Sciences de l'Institut de France and RSC - natural sciences
- Ukrainian Academy of Sciences and RSC - natural sciences.

3. The Canadian Institute for Advanced Research (CIAR) [3]

In today's highly competitive global economy, a nation's prosperity is increasingly linked to its ability to develop and apply knowledge. This requires not only a greater investment in science and technology, but also in the creation of new and flexible research structures that are able to adapt quickly and cost-effectively in a dramatically changing world environment.

The Canadian Institute for Advanced Research is a private, non-profit corporation established (1981) to focus both intellectual and financial resources on research that is critically important to our future.

The Institute invests in people, not in buildings and equipment. It pays the salaries and interaction costs of outstanding researchers both in Canada and abroad. By freeing these talented individuals from routine administrative and teaching tasks at their base institutions, CIAR has created a one-of-a-kind scientific network that makes the best possible use of every research dollar.

CIAR's programs focus on three principal areas:

- frontiers of modern research
- foundations of emerging technologies
- societies' adaptation to change

Within the first five years, the CIAR began programs in

- artificial intelligence and robotics
- cosmology
- evolutionary biology
- population health
- law and the determinants of social order

To these have been added programs in

- superconductivity
- economic growth and policy
- earth system evolution
- human development
- soft surfaces and interfaces

For each of these programs, the potential topic was first identified by the CIAR Research Council, which regularly investigates high priority areas of research whose scope may be difficult to address within existing institutions. Following the Research Council's initial examination, a task force of experts is usually formed to explore the challenges of the particular field. If the task force's report is accepted by the Council, the President then works with an "implementation group" to find a distinctive focus for the program and to identify potential Fellows. Each program is reviewed in the fourth year by an external panel of international experts who are asked to make a recommendation on whether the program is to continue or not.

By creating a network of outstanding researchers at many different institutions in Canada and abroad, CIAR brings together a critical mass of talent to confront these complex issues in a collaborative and constructive fashion. Today, the Institute's programs in these areas involve more than 180 researchers based mainly in Canada and about 15 countries. CIAR raises about $10 million annually in support of these programs, from a wide cross-section of individual donors, private industry, foundations, provincial governments, and the Government of Canada.

4. Ontario Centres of Excellence [4]

The Ontario Centres of Excellence Program was inaugurated by the Province of Ontario in 1988, with a grant of $200 million for five years. Seven Centres were selected by an international panel from 28 submitted proposals. Each proposal was evaluated against the following criteria:

- the quality of the researchers
- the quality of the research program
- the potential for Ontario to develop a lead position in the program area, given the existing industrial basis in Canada and internationally.

The expectations were that each Centre would be:

- first and foremost a centre of advanced research and study with a long-term perspective
- built upon existing research excellence and areas of demonstrated strength and potential strength
- dedicated to long-term research with the expectation of maximum potential benefit to the economy of Ontario. It should ensure that academic researchers are aware of industry's needs, and that industry is aware of the knowledge available through academic researchers.

The seven Centres are:

- the Institute for Space and Terrestrial Science
- the Information Technology Research Centre
- the Manufacturing Research Corporation of Ontario
- the Ontario Centre for Materials Research
- the Ontario Laser and Lightwave Research Centre
- the Telecommunications Research Institute of Ontario
- Waterloo Centre for Groundwater Research

Each Centre has mounted intensive programs in their specialized fields of advanced research and applications. In addition, each is fulfilling its mandate in training world-class researchers, encouraging the transfer and diffusion of technical knowledge to industry, and developing strong links with Ontario Industry. Joint efforts in research, communications, education, and interaction with industry have been nurtured. Thus, by working together, the Centres are adding important new strengths to the overall program, and ensuring that the total effort is immeasurably greater than the initial expectations.

In the eight years of their existence, the Centres have been supported by three different governments. After the first five years, and demanding reviews by international and industrial panels, they have been judged to be fulfilling their mandates in every respect, and continue to be funded by the province of Ontario, with growing financial support by industry.

5. Networks of Centres of Excellence of Canada (NCE) [5]

The Networks of Centres of Excellence Program, launched in 1990 by the Government of Canada, is an intensive national R&D team effort aimed at enhancing Canada's industrial competitiveness and social well-being in a new global economy. The program draws together many of the country's top researchers from universities, industry, and government to work in partnership on research challenges and opportunities that are vital to Canada's future prosperity and quality of life. This Program was initiated with a grant of $240 million for five years.

Following a rigorous peer review involving international experts, fifteen NCEs were chosen out of 158 applications, to meet the following specific objectives:

- to stimulate leading-edge fundamental and long-term applied research of importance to Canada
- to train and retain world-class scientists and engineers in fields that are critical to Canada's industrial competitiveness and quality of life
- to integrate excellent Canadian research and technology development efforts into national networks with the participation and partnership of universities, private sector, federal government and the provinces
- to develop strong university-industry partnerships in order to accelerate the diffusion of advanced knowledge to industry.

After the first five years and peer reviews, ten of the NCEs were recommended for renewal, and four new networks were added. These fourteen NCEs form a nationwide network pursuing leading-edge research in areas of strategic importance to Canada, and working with industry to create commercial opportunities out of the results. The present NCEs are:

- Intelligent Sensing for Innovative Structures
- Institute for Robotics and Intelligent Systems
- Canadian Institute for Telecommunications Research
- Micronet-Microelectronic Devices, Circuits and Systems
- Mechanical Wood-Pulps Network
- Sustainable Forest Management
- Concrete Canada
- Canadian Bacterial Diseases Network
- Canadian Genetic Diseases Network
- Neuroscience Network
- Protein Engineering Network
- Health Evidence Application and Linkage Network
- Inspiraplex
- Telelearning Research Network

A total of 629 organizations from universities and the public and private sectors are involved in these networks. These include 405 companies, 76 government departments and agencies, 37 hospitals, 48 universities, and 63 other organizations, employing over 4,000 researchers, students, and highly qualified personnel.

6. Summary

The activities of the Academy of Science of the Royal Society of Canada are purposely of a scholarly nature, dealing primarily with studies of current issues of local, national, and international interest. Over the years, the Society has been asked for advice by the Federal Government and its agencies, and continues to act in this capacity. It makes recommendations for senior appointments to national posts, and comments on government policies.

While the roles of the Canadian Institute for Advanced Research, the Ontario Centres of Excellence, and the Networks of Centres of Excellence of Canada, vary in objectives and content of selected research, they all focus on the collaboration of researchers in academia, industry, and government. The separation of research into these three areas has been a long-standing problem in Canada, and perhaps in other countries, but much has been accomplished in the past decade with the establishment of these programs. Also, cooperation among Canadian scientists and engineers has increased, and between them and scientists and engineers in major universities in the United States, Europe, and Asia.

146

One of the most important benefits of these programs has been in the retention of intellectual talent in Canada, and in enticing the return of highly-trained Canadian researchers. Other worthy achievements have been the building of high-quality research programs in frontier areas that have potential economic, social, and intellectual importance. These programs have had direct impact on education in schools, in universities, and in industry, on environmental issues, on health and welfare, and in industrial development.

The establishment of these programs of cooperation and networking of scientists and engineers has been of crucial importance to researchers in Canada in this difficult period when provincial governments and the federal government are down-sizing and retrenching. There is little doubt that with continuing support of research, benefits will accrue to industry, to the economy, to health, and to our social well-being.

References

1.	Collins, P.M.D. (1996) Science International Newsletter, No. 61, p. 16.
2.	Annual Reports of The Royal Society of Canada, Ottawa, Canada (1994, 1995).
3.	Annual Reports of the Canadian Institute for Advanced Research, Toronto, Canada (1994, 1995).
4.	Annual Report of the Ontario Centres of Excellence, Toronto, Canada, (1994, 1995).
5.	Annual Report, Networks of Centres of Excellence, Ottawa, Canada, (1996).

TRANSITION PERIOD PROBLEMS OF THE NATIONAL ACADEMY OF SCIENCES OF THE REPUBLIC OF ARMENIA

FADEY SARGSYAN
President of the National Academy of Sciences of Armenia
24 Bagramian Ave., 375019 Yerevan, Armenia

1. Introduction

The present state of the National Academy of Sciences of Armenia is characterized by isolation from the former Soviet Union Academy complex, interruption of centralized research orders and distortion of links between the Academies. The budget assigned for science has been relatively reduced, and the institutional and program funding has been interrupted. As a result, scientific potential tends to decrease. Migration of outstanding scientists abroad is being observed, difficulties to involve new young researchers in science are obvious, and the "aging" of scientific schools is becoming apparent. All these things are happening under existing unfavorable conditions, such as an economic crisis, informational isolation, deplorable conditions of the infrastructure (buildings, laboratories), and provision of materials and equipment. In order to maintain and, what is more important, to develop science in Armenia, the following cardinal problems should be solved:

- Integration with the international scientific community, participation in international scientific programs and associations. The International Association of Academies can play a great role in solving these problems by integrating the Academies of the CIS countries which will result in the formation of a common scientific environment and the restoration of the former relations on the basis of the new economic and political conditions.
- Working out of a long term science policy, the most important part of which should be the definition of priorities in science and technology, the diversity of funding channels and mechanisms to promote the development of scientific directions and innovative research. The government should define the targets for the state R&D programs. Also, it is necessary to put into practice the results of the scientific discoveries

Ch. Proukakis and N. Katsaros (eds.),
The New Role of the Academies of Sciences in the Balkan Countries, 147–153.
© 1997 *Kluwer Academic Publishers. Printed in the Netherlands.*

and to use the raw materials effectively by means of modern technologies.

- Development of legislation on science and technology, establishing of favorable taxing and credit policies.
- Creation of integrated structures of science and education, interdisciplinary scientific units to stimulate research and development of activities on the border line of different directions.

2. Long-Term Science Policy and Integration with the International Scientific Community

The present situation of practical isolation of the Academies in the FSU republics affected especially the development of S&T in the Republic of Armenia. It is obvious that the Republic has no possibilities to finance research programs as before. Meanwhile, we are well aware of the fact that there is no concept of national science policy as such, and we have no right to lose the existing scientific potential of the Republic created during the centuries and which is, to our mind, our common achievement.

Being in a very difficult and financially strained situation for creative scientific research, during the last years, Armenian scientists have made great efforts, on one side, to contribute feasibly to the solution of the arising complicated problems in the Republic connected with transition period economic development, and on the other side, to rehabilitate and to further necessary scientific ties with their partners. In this direction, considerable work has been done by the International Association of the Academy of Sciences and we try to take an active part in its activities. Here, I want to mention a very positive and principal contribution of the President of the Association, Academician B. E. Paton, directed to the consolidation of our efforts. Undoubtedly, the Academy of Sciences of the Russian Federation remains our main partner with whom we have joint scientific organizations, international centers, etc. The Academy of Sciences of the Russian Federation shows possible assistance in all our problems. Science is a great property of our country, an integral part of a high intellectual potential of our nation. A good many talented individuals distinguished in the fields of science, culture, arts, politics, and economy have originated from Armenia. Consequently, preservation and development of created scientific potential in conditions of independence is a priority task. Therefore, together with the implementation of well-grounded reforms directed to the realization of important republican projects, we make efforts to use more efficiently the possibilities of collaboration with different organizations of foreign countries, as well as those rendered by international scientific foundations. At the same time, it is principal to preserve the scientific potential in the field of basic research. Giving preference to scientific and technological elaborations, we bear in mind that applied fields of science can not survive and develop without basic research. Loss of potential in the field of basic research would have been an unrecoverable and unpardonable mistake, since it serves as a basis for technological and more general, applied sciences. We have also to mention the importance of fundamental, theoretical elaborations for university science and education.

We are deeply convinced that the results of fundamental research carried out in the Institutes of the National Academy of Sciences of the Republic of Armenia can serve as a basis for the creation of quite competitive technologies in the international market. Several projects carried out by our scientists abroad come to prove this.

One of the inevitable problems of a transition period is the down scaling of state supported projects and, consequently, of scientific staff. This leads to considerable social problems affecting vital interests of an individual and to a danger of dissolution of established scientific schools, directions created during decades and having national significance. Previously, the volume of funding scientific projects was determined according to all-union scales, and since our Republic had a high scientific potential, huge scientific centers were established in the country, involving a great number of specialists. It is natural that this number should decrease. However, two important questions arise here. The first concerns the selection of the best scientific staff for the preservation of the scientific potential. The second question has a social coloring, that is what to do with those who are not chosen? What will be their further career? At present, this group needs definitely some social protection. Most of them have worked honestly all their life, and many did not reach retirement age. It is also too late to be trained for a new profession. Meanwhile, aging of the research staff is taking place, so the most talented, young scientists leave the institutes. Of course, taking into account the existing reality, we simultaneously reorganize the institutional structure, aiming to preserve the most vital directions. However, we are not always successful in our efforts. Therefore, more radical measures are necessary, otherwise, the scientific potential of the Republic will suffer unrecoverable qualitative and quantitative damage.

It is also necessary to mention that when it concerns the scientific work force we should not proceed only from the necessity of staff reduction. Thus, for instance, according to preliminary results of peer evaluation for this year, from 13 projects submitted by the Institute of Zoology, including 103 researchers, only 1 project including 9 researchers was approved. The Institute must have practically closed. And this is the case when recently the Russian Academy of Sciences created the Scientific Center of Applied Zoology on the basis of this Institute, highly estimating its scientific achievements. The threat of shutting down the Institute was partially staved off by the Council of Science and Advanced Technologies at the Prime Minister of Armenia. The same took place with the Institute of Geophysics in Gyumri, Institute of Hydroecology in Sevan, etc. The aforementioned is indicative of the inadmissibility of the inadequate and formal approach to the question of science development. To avoid such cases we have to improve radically the existing system of project evaluation and funding. Firstly, project funding should be envisaged as a separate item of the science budget. Besides such financing, it is also necessary to start stable institutional funding of the institutes directed to financing of basic research and the development and maintenance of an experimental base and science infrastructure.

The volume of institutional funding should be determined for every field of science separately and constitute at least 25% of the total NAS budget. We have several unique scientific labs, the preservation of which is of exceptional importance for international science. These are the Byurakan Observatory, the Institutes of Zoology, Botanics, Physics, a rich scientific and technological library, the Institute of Genocide, etc. It requires special investments as well.

It is also a task of primary importance for the NAS to work out well-grounded proposals on the principles of further development of the national science and technology system which should lay on the basis of research funding policies.

The existing peer evaluation system needs to be seriously revised. In the expertise, more importance should be given to the scientific councils of the Institutes and Departments of the Academy where leading exports of the country are involved. We think that the methods applied in the present peer reviewing process do not adequately respond to the structural organization of the institutes. Staff reductions which have been already realized more than once are also inevitable. But these questions should find their solution in the research institutes themselves. The peer evaluation of the projects developed in the institutions must be carried out according to the methodology and procedures designed and developed by the Academy itself, based on the analysis of progressive international experience.

At the same time, the government should announce competitions and allocate a certain number of grants for the purpose of solving priority problems, the immediate solutions of which are of crucial significance.

The expertise of projects submitted for evaluation should be carried out under the aegis of coordinating the state body.

To summarize the above-mentioned, we see three constituent channels of research funding, namely: institutional, project-based and program-based.

The Academy is preparing a plan of radical reorganization of its institutional structure.

First of all, scientists must concentrate their efforts on the issues of the sphere of their own expertise, as well as on the problems competitive at the international science market and based on the state priority interests. Otherwise, science trends can not be developed and the obtained results, lacking any implementation possibilities, will serve as an obstacle for their further development. It is necessary to utilize most effectively the possibilities rendered by international foundations.

Thus, the possible solution of the problem may be, on the one hand, an increase of state subsidies and intensive utilization of grant possibilities rendered by international foundations, and, on the other hand, a correct determination of scientific development, adequately organized project expertise, and radical improvement of scientific and institutional activities.

Another essential problem for our Academy is the aging of research staff, the outflow of young specialists and the difficulties connected with training the new generation of scientists.

We have initiated the organization of higher advanced academic courses at the Academy for talented young scientists. There are favourable conditions in Armenia for the creation of scientific and educational units, and the Academy is currently considering corresponding proposals.

In modern conditions, one of the most urgent problems is the creation of a data exchange network. In spite of evident progress, we have no possibility to provide all the institutes of the Academy with access to the global system of Internet.

The existing financial difficulties do not allow us to realize the exchange of scientists and literature. But more serious is the problem when, due to the same reasons, we have no possibility to publish the scientific work of our scientists, hundreds of which wait for their turn in printing houses and on the shelves losing their topicality.

In present conditions, the National Academy of Sciences is making efforts to establish effective international collaboration. Our Academy joined the International Council of Scientific Unions (ICSU) and the Interacademy Panel on International Issues.

During the last years, the institutes of the Academy joined different International Scientific Associations and Unions, namely: the International Federation for Information Processing (IFIP), the International Brain Research Organization (IBRO), the International Astronomical Union, etc.

Our Academy is working intensively towards the integration to the international scientific community, and the utilization of opportunities made available by various international and national scientific organizations and foundations, particularly of such projects of the European Community as COPERNICUS, TACIS, INTAS, PIKO, etc.

Our institutions are currently involved in different joint projects with their partners.

A substantial role is being given to our active participation in the scientific and technological programs of the North-Atlantic Union. Our participation at the present arrangement conducted under the auspices of NATO is also indicative of this.

We consider this forum the most considerable step towards our integration in international science.

We consider as important the establishment of bilateral co-operation with national scientific organizations and institutions of different countries - donors of science. This process is being promoted by the representations and embassies of these countries in Armenia. In particular, the French Embassy in Armenia has greatly contributed to the establishment of collaboration between the French and Armenian scientists in astrophysics and astronomy, sponsored by the French government. Similar work is being conducted with other countries as well. We signed the Memorandum of Understanding with the Royal Society of Great Britain, which has been in progress successfully for the last two years. Joint research in Information Science is conducted within the framework of collaboration with the German Scientific Research Society. We are working successfully in collaboration with the University of California on a program of inculcating fast growing trees. Preliminary negotiations are also being carried out with the organizations of such countries as the USA, Portugal, Greece, Japan, China etc. In spite of the transition period and financial problems encountered by both countries, we continue our cooperation with the Hungarian Academy of Science, started in the early 60s.

Our scientists greatly appreciate the possibilities made available by the Soros Foundation.

We have to mention also a great contribution of other foundations and organizations in the support and promotion of science, namely: the International Research and Exchange Board, the Fullbright Foundation (USA), the Humboldt Foundation (Germany), the Gulbenkyan Foundation (Portugal), etc.

In spite of a difficult complex financial situation, our work for integration in the world science will be continued and extended according to all the trends mentioned, taking into account the modern realities.

At the same time, we proceed from the fact that the newly obtained independence of our republic creates new possibilities for our scientists to participate in international scientific programs, especially with scientific organizations of CIS representing a common scientific and technological area.

The main aspect in this direction is the elaboration of a well-grounded action plan, the determination of scientific and technological priorities, possible sources of financing, securing science and invention development. Undoubtedly, this work will be tightly connected with present requirements and peculiarities of the Republic. To our mind, such priority directions may be the following:

- creation of a scientific informational background for the solution of applied tasks on the basis of further development of basic research (mathematics, physics, biology, chemistry);
- investigation of new fuel and energy sources;
- environmental studies, including biotechnology, stress investigations, mineral extraction, waste recycling and other ecological problems;
- Armenian studies.

All this requires a decisive step towards the implementation of the results of scientific investigations.

One of our most important tasks is to complete the elaboration and submit to the National Assembly the law on science and NAS. The first version of such a law has already been worked out. The law should ascertain particularly the place and role of the NAS, ways of financing, including subsidies from the state budget, realty possession right, etc.

We hope that the aforementioned law will allow us to some extent to relieve our condition in question of taxation, to improve social conditions of scientists, to make a distinction in the existing diversity of scientific awards and degrees, etc.

THE ACADEMY'S ROLE IN PROMOTING INTERDISCIPLINARY SCIENTIFIC RESEARCH

MALCOLM JEEVES
President, Royal Society of Edinburgh
22-24 George Street, Edinburgh, EH2 2PQ, United Kingdom

1. Introduction

When the Royal Society of Edinburgh was constituted in 1783, science was relatively circumscribed, and scholars could meet regularly and usually converse freely on the basis of an assumed, shared body of scientific and literary knowledge. The ease with which one could move from discipline to discipline is well illustrated by the experience of an early fellow of the Society, William Cullen (1710-1790). He moved smoothly from being Professor of <u>Medicine</u> at Glasgow University to being Professor of <u>Chemistry</u> and <u>Physics</u> at Edinburgh University. When the Society was formed, it was sufficient to divide it into a <u>literary class</u> with 93 members, and a <u>physical class</u> with 72 members. The literary class included men like the economist Adam Smith and the physical class, Joseph Black, James Hutton of the 'theory of the earth' fame, James Watt the engineer, and James Gregory the mathematician. The lectures and discussions of the Society were published in the Society's in-house publications, the <u>Transactions</u> and <u>Proceedings</u>.

Whilst the problems that arise at the interface of the so-called two cultures, the humanities and the sciences, continue to be widely discussed, there are similar problems which arise in maintaining adequate communication between different branches within the natural sciences. The possible effects of uncalled for separation between scientific disciplines are potentially serious when one recalls that any examination of the history of science indicates how frequently important discoveries have been made at the <u>interface</u> of different scientific disciplines, rather than in the centre of any one of them. <u>For this reason alone, the academy today has an important, indeed I would argue a crucial role, to maintain a strong interdisciplinary emphasis at a time when science continues to fragment into ever smaller specializations and sub-disciplines.</u>

The Society's membership is drawn from the whole field of learning. Fourteen specialist committees select candidates for membership from the

Ch. Proukakis and N. Katsaros (eds.),
The New Role of the Academies of Sciences in the Balkan Countries, 155–160.
© 1997 *Kluwer Academic Publishers. Printed in the Netherlands.*

156

different disciplines of science, medicine, the arts and humanities, law, education, business and administration. Their interests are catered for by specialist meetings and symposia. But, perhaps even more importantly, the Society pursues a deliberate policy of addressing complex and often controversial issues. These often relate to different aspects of social or environmental policy. By their very nature they demand a many-sided, inter-disciplinary approach and treatment. Because of the range of skills and professional experience vested in the Society's membership it is uniquely competent to organize meetings and conferences which invite participation by specialists from many different fields of research, as a regular part of its programme of activities.

Complex issues, especially those which relate to society, often have a political content and feelings may run high. By providing a neutral setting, without any kind of prior political commitment, the Society can play the role of 'honest broker'. Intellectual adversaries can be brought together in an effort to find common ground for a better understanding of different points of view. Such occasions serve to enlarge intellectual horizons which may have become constrained by over concern for special interests.

A few examples of meetings and conferences held during recent years will illustrate the approach. The first two examples refer to biological and medical issues which raise moral and ethical questions. The next two are rather different. They focus on the national implications of developments in science and engineering technology. The last two are international in scope and deal respectively with the reaction of biological systems to implanted materials and the global, environmental problems posed by acid rain.

1.1. BIOMEDICAL/ETHICAL TOPICS

(i) In 1993, the Royal Society of Edinburgh mounted a two day international meeting under the title **'To Treat or Not to Treat?: Dilemmas Posed by the Hopelessly Ill'** (Table 1). The medical issues were introduced by a Professor of Neurosurgery from Glasgow, the ethical issues by a Professor of Philosophy from Glasgow and the legal issues by a Professor of Law from the Free University of Berlin. The programme included discussion of issues such as handling acute crises, (discussed by an anesthetist), the effects of brain-damage (by a mental health specialist), and the particular problems of children and the elderly by medical specialists in those areas. On the second day of the meeting, discussion widened to benefit from experiences overseas, and included contributions concerning medical decisions at the end of life, based upon the Netherlands experience, and experience in the United States of America. I believe that this kind of breadth of treatment, as well as its depth, could not have been adequately achieved from <u>within</u> the narrow confines of one medical or biological discipline. By bringing together this range of specialists, it was possible, for example, to focus on the inappropriate use of treatments when a reasonable recovery is not possible, or when prolonging a life appears not to be in the patients best interest. It was recognized that the ethical issues raised are often troubling to doctors as well as worrying to the public. In addition there can be legal implications, particularly if the patient is not competent to express preferences about his or her management. Such a conference, bringing together doctors, nurses, lawyers, philosophers, social scientists and members of the general public was thus timely and enlightening.

(ii) My second example of a multi-disciplinary kind, this time bringing together biomedical specialists and engineers, was a two day meeting held in 1994 under the title **'Technology for Physically and Mentally Handicapped People'**. On this occasion, specialists from rehabilitation medicine, and from those specializing in the development of communication aids for language and learning, came together for two days. Specialists in the development of future communication technology, of rehabilitation engineering services, and experts in electrical implant techniques addressed problems of common interest. Again the international dimension was represented by speakers from overseas. In this category, had time allowed, I could have included a recent successful meeting on Gene Therapy (some details are given in Table 4).

1.2. CONFERENCES ADDRESSING NATIONAL ISSUES IN SCIENCE AND ENGINEERING

My second group of topics is illustrated by two meetings held in recent years to consider first, **'The Science Base: Underpinning the Future in Scotland'** and the second, held in 1990, under the heading **'Engineering: Scotland's Future'** (Table 2).

(i) The 1994 meeting was partly in response to a Government White Paper on the organization and funding of scientific research. In the special context of Scotland, the conference benefited from the contributions of the Head of Research of one of the largest pharmaceutical firms in Great Britain, as well as from the Director of Funding from the Scottish Higher Education Funding Council. The presentations of these two speakers were then supplemented by a round table discussion, bringing in the viewpoints of particular groups from industry such as, for example, chemicals, Atlantic and North Sea explorations, agriculture, electronics, engineering and the financing of new business (by the Governor of the Bank of Scotland). These topics were given a complementary perspective by a group of presentations under the heading of 'Science and Society'. Topics covered

158

by specialists in this area included discussion of the natural resources in Scotland and the research requirements thereof, the future of the science base, and in particular, the role of secondary education, and the Scottish Higher Education Funding Council and the funding of the science base. Thus, an issue of national interest was given an in-depth airing with broad perspectives in a single meeting at the Royal Society of Edinburgh.

(ii) The other example I have taken focuses more on engineering than on basic science. At a two day meeting, held in May 1990, to mark the centenary of the opening of the Forth Bridge, a gathering of distinguished engineers both in academia and in industry presented a wide ranging overview of the current situation of engineering in Scotland, and looked ahead to its future. Topics presented and discussed included information technology, manufacturing industry, energy, North Sea developments, transport and development, both regional and national perspectives, and the wider background of the Scottish economy, as well as the civil engineering infrastructure in Scotland. Thus once again the diversity of perspectives were focused on a common problem, followed on the second day by site visits to some of the largest electronics and engineering establishments in Scotland, including Torness Nuclear Power Station.

1.3. SCIENTIFIC ISSUES WITH AN INTERNATIONAL DIMENSION

Whilst some of the issues discussed above could legitimately be described as, to a degree, parochial because they were concerned with relatively localized issues within Scotland, other issues which span national frontiers have also figured prominently in the discussions and conferences of the Royal Society of Edinburgh in the past decade (see Table 3).

(i) The first example was one of a series of seven yearly conferences which the Society undertook to arrange in 1990. It lasted six days on the topic **'Acidic Deposition - It's Nature and Impacts'**. The conference focused on nine aspects of the nature and impacts of atmospheric pollutants, including ozone, whose occurrence is dependent upon the combustion of fossil fuels. It included critical assessments of many national and international programmes of pollution research and was arranged around a series of plenary and concurrent sessions. Since the meeting lasted six days, I can here give no more than the titles of a subset of the plenary lectures to illustrate the range of topics covered. These included a lecture on free radical chemistry of surface air by Dr. Ehhalt from Germany; on direct and indirect effects of gaseous pollutants on enzymes, by Dr. Anderson from USA; on critical physiological changes in polluted plants by Dr. Mansfield of the UK; on the extent and rate of surface water acidification, comparison of measured and predictive evidence by Dr.

Wright from Norway; and on fresh water acidification, reversibility and recovery, comparisons of experimental and non-manipulative systems by Dr. Schindler from Canada. At the same time, there were concurrent sessions on nine fields of interest, ranging from chemistry of atmospheric pollutants, through the physiology of plant responses to pollutants, to atmospheric pollutants in forests and including the effects of pollutants on fresh water plants and animals. Thus the range and depth of topics covered are such as would be unlikely to be handled by the meetings of specialists from <u>within</u> one scientific discipline only, <u>hence the importance of the interdisciplinary role of the academy</u>.

(ii)Reactions to implantable materials. In this meeting the Society brought together surgeons, physicians, biologists, engineers, dentists and others to address the practical difficulties met with in their work. Scientists familiar with the mechanisms of bacterial corrosion in off-shore oil rigs exchanged experience and opinion with physicians tackling the problems of new haemo-compatible materials, with dentists working with dental biomaterials and with orthopaedic surgeons who are handling a range of artificial substitutes for bone.

1.3. DISCUSSION

I have suggested that one of the key roles of the academy is to bring together specialists not only from within diverse scientific disciplines but also spanning the humanities, all of whose contributions are vitally important for the comprehensive discussion of contemporary issues, particularly in the biomedical area. Groups that may otherwise remain out of communication with one another may, at the initiation of the academy, be brought together, and helped to recognize the important interfaces and common interests between their respective specialist knowledge. It seems highly likely that with the accelerating pace of research in, for example, gene therapy (see Table 4), there will be issues which cannot be confined to the biomedical arena, but spill over into ethical issues of considerable importance for human welfare, both individually and socially. Another area in which social/ethical/moral issues arise is psychopharmacology. There have already been examples, in the United States of America, of the 'cosmetic' use of antidepressant drugs, such as Prozac.

The academy's role in facilitating interdisciplinary discussions will become increasingly important if the benefits of developments in science and engineering are to be applied to problems of national interest in terms of social and economic development. We have given two examples of this above, within the Scottish scene. In both cases, highly successful meetings were held.

Finally, the academy has an important role to play on the international scene, where the issues at stake, scientifically and environmentally, are too large to be handled by one area of the world, and where a proper understanding of the dimensions of the problem only becomes apparent from the contributions of countries widely spread geographically. In this context, we have considered the example of the ongoing problem of acidic deposition. In this realm, it is possible for scientists to come together and to consider issues on which governments, both nationally and internationally, are seeking advice. These are complex issues, in terms of their scientific, economic, ethical and political ramifications, as well as often being issues widely debated in the public domain. On such issues, the Society's role of 'honest broker' can be a key factor in promoting rational discussion since, as a Society, it is not committed to any political or ideological stance.

BIOMEDICAL / ETHICAL TOPICS

TOPIC	CONTRIBUTING DISCIPLINES
To treat or not to treat (1993)	Neurosurgery, Anaesthetics, AIDS research, Mental Health, Child Health, Medical Ethics, Law, Sociology, Philosophy
Technology for Physically and Mentally Handicapped People (1994)	Psychology, Communication technology, Rehabilitation Medicine, Orthotics, Health Science, Rehabilitation Engineering, Medicine, Neurology, Bioengineering, Nursing studies, Computer science, Occupational therapy, Physiotherapy, Audiology, Geriatrics

Table 1

NATIONAL ISSUES IN SCIENCE AND ENGINEERING

TOPIC	CONTRIBUTING DISCIPLINES
Engineering Scotland's Future (1990)	Industrial engineering, Policy studies, Mechanical engineering, Civil engineers, Economics, Shipbuilding
The Science Base: Underpinning the future in Scotland (1994)	Pharmaceutical Research, Business and Marketing, Basic Skills of Fermentation and Distillation, Chemical Industry, Atlantic and North Sea Exploration, Agriculture, Banking, Electronics Industry, Civil Engineering, National Resources Research, Higher Education - policy and planning, Secondary Education - policy and planning, University Administration

Table 2

SCIENTIFIC ISSUES WITH AN INTERNATIONAL DIMENSION

TOPIC	CONTRIBUTING DISCIPLINES
Reactions to Implantable Materials (1990)	Surgeons, Physicians, Molecular and Cell-biology, Dental biomaterials, Metallurgy and Materials, Biology, Allergists, Protein and Molecular Biology, Bioengineering, Pure and Applied Chemistry
Acidic Deposition (1990)	Chemistry, Economics, Plant physiology, Forestry, Soil Science, Materials Science, Hydrology, Zoology, Fishery research

Table 3

A CONTEMPORARY ISSUE

TOPIC	SUB-TOPICS	CONTRIBUTING DISCIPLINES
Gene Therapy (1995)	Technical challenge	Molecular genetics, Veterinary pathology
	Clinical opportunities	Virology, Molecular pathology, Paediatric oncology
	Ethics and social responsibilities	Eukaryotic molecular genetics, psychology and genetics, human genetics, gene therapy advisory committee
	Industrial and Health care perspectives	Pharmaceutical Research companies, Scottish Home and Health Department

Table 4

THE ROLE OF ACADEMIES OF SCIENCES IN PERIODS OF TRANSITION

HENNING SØRENSEN
The Royal Danish Academy of Sciences and Letters
H.C.Andersens Boulevard 35, DK-1553 Copenhagen K

Abstract: Academies of sciences are learned societies, which have different roles to play in different countries. In the research system of the former Soviet Union and in many Central and East European countries, the academies, in addition to being learned societies, also were responsible for the administration of en extensive system of research institutes. The break up of the Soviet Union and the accompanying transition from a centrally planned to a market oriented economy resulted in severe problems for the survival and development of the research system of these countries. These problems are elucidated by means of the situation in the three Baltic countries, Estonia, Latvia and Lithuania, with special emphasis on Latvia. Research-based industries practically ceased to exist after the links to the Soviet Union were broken, and there was a sharp decline in the appropriations for research. In Latvia, the Academy was transformed into a learned society without responsibility for administration of institutes, a research council is now responsible for distribution of funding of research, and an integration of research institutes and universities has been initiated.

"The academies of sciences are the conscience of the scientific world, or rather the scientific conscience of the world itself"; this was the characterization of science academies given by Professor Jacquinot, the former president of the Académie des Sciences de France, at an international academy conference in 1982. This states in a few words how science academies should act, and why they are important institutions. It also indicates that academies not only work within a national framework, but that they also have an international, or rather global role to play.

In this contribution, I have been asked to discuss the role of academies of sciences in periods of transition; that is when countries are in transition from one political and/or economic system to another, as exemplified by the countries of the former Soviet Union and many developing countries. This generally results in decreasing financial resources to research, which, coupled

Ch. Proukakis and N. Katsaros (eds.),
The New Role of the Academies of Sciences in the Balkan Countries, 161–170.
© *1997 Kluwer Academic Publishers. Printed in the Netherlands.*

with a dwindling public interest in, especially, the basic sciences, leaves the research systems, including the academies, in an exposed position. How will this affect the research systems and the academies, and what can the academies do to help the countries through such difficult periods?

First, a few general statements about academies of sciences. Their purpose is to promote science and scholarship, and in many academies also to further the transfer of scientific achievements to applications in society. They do this in their capacity of learned societies counting as members distinguished scientists and scholars. Their strength is the scientific status and international networks of the members. Besides, many academies are old and keep in their libraries and archives material of great scientific and general importance. They may be termed depositories of knowledge and culture.

Many academies cover the whole spectrum of fundamental sciences, from mathematics and natural sciences over the humanities to the social sciences. Some academies even include the applied sciences. In many countries, however, the natural sciences, the humanities-social sciences and the technical and medical sciences have each their academy.

The furthering of interdisplinary research is often stated as one of the main goals of academies. This is, clearly, most easy in the academies that cover the whole spectrum of fundamental research. In these cases, the academies serve as meeting places for researchers from many scientific disciplines, and are in many countries the only places for such meetings of top scientists. This should be counted as one of the strong sides of the traditional academies.

Interdisciplinarity is more difficult to practice in the countries that have separate academies in the major sectors of research, but should nevertheless be an important activity for all academies, since interesting scientific developments often take place in the boundary fields between the major disciplines of research. This is especially important in countries in transition, in which it is necessary to apply a holistic view on the whole system of higher education and research in order to secure the balance between disciplines, the conditions for developing new research along their boundaries, and the transfer of results to society. The authorities may have difficulties in considering all these aspects when they have to solve the urgent problems in their countries. Similarly, the universities and research institutions, that are the participants in and perhaps victims of the transition processes, each fight for their own existence and development. Superior, impartial advice is clearly needed; the academies of sciences are institutions that should take up this responsibility, and in many countries already do so.

The traditional academies that function mainly as learned societies can most easily meet these demands and should do it. There are, however, academies of sciences, which are themselves parties in the transition processes, namely those that play the role as research councils and even ministries of research. Such academies may have difficulties in giving impartial advice. They direct research institutions and centres, institutions that may no longer correspond to the needs of the surrounding society. Who can give the impartial advice in such cases? International organizations such as the OECD, the European Science Foundation (ESF), and other countries could be asked to give advice about the future development of the research system. I can as an example mention the evaluation of the research systems of the three Baltic states, Estonia, Latvia and Lithuania, undertaken by the research councils and academies of Sweden [1], Denmark [2] and Norway [3]. I was involved in the evaluation of Latvia and shall base my presentation on the experience obtained in the Baltic states, and especially in Latvia.

The research systems of the three Baltic states were formerly part of the Soviet organization headed by the Academy of Sciences of the USSR. The staffs of many of the research institutes of the Baltic countries counted hundreds of scientists and a number of the institutes were at a very high international level, as was documented in the above-mentioned Nordic evaluation of the research systems of the Baltic states. Research institutes and higher education were separated. The large research institutes were part of big networks involving military, space and industrial research. These networks disintegrated with the break up of the Soviet Union. The three countries were left with a number of powerful research institutes that were clearly much too big for their needs and economic capacity. Furthermore, the research-based Soviet industries, that were located in the Baltic states, practically ceased to exist after 1990. The ensuing economic contraction and a lack of hard currency make the maintenance of the research capacity, including staff, scientific equipment, libraries, etc., very difficult and there is a great risk that the infrastructure of the research system will deteriorate, unless other countries or international organizations offer their assistance.

The Baltic states have taken a number of measures in order to preserve and develop their research potential and to stimulate the active involvement of researchers in solving current economic, social and cultural problems and in reconstructuring the industrial system.

The economic contraction in these countries has resulted in a sharp decline in the appropriations for research. The gross domestic expenditure on research varies from 0.45 to 0.7 %. This has resulted in a decrease in the number of tenured staff to less than half of the figure for 1991. Furthermore, salaries are lower than the existence minimum, which forces researchers to take more jobs in order to survive, or to leave to better paid positions within the countries or in other countries. There is, thus, a severe internal and external brain drain. But it should be emphasized that there is in spite of this a very strong will among researchers not only to survive but also to develop their research. The low salaries do not, however, favour the recruitment of young scientists.

It has not yet been possible to build up a new research-based industry; less than 10 % of the expenses to research are covered by industrial enterprises. The state budget remains the main source of financing of research and higher education.

These harsh conditions are reflected in the activities of the academies of sciences of the Baltic states. I shall elucidate this using the Latvian Academy of Sciences as an example. It was a powerful organization in the Soviet Union heading a number of very big and, in some cases, very advanced research institutes at the highest international level. This was, of course, difficult to accommodate in a country with 2.6 million inhabitants. It was therefore decided in July 1990, only two months after the declaration of independence, to establish a National Research Council that was given the superior responsibility for the distribution of funds to the various research sectors, including the academy institutes. A further step was to change the Academy of Science into a society of elected members. It no longer directs research institutes; these are now supported by the Research Council, and an integration of research institutes with the universities has been initiated. One can say that the former top-down system has now been replaced by a more bottom-up system, where the distribution of funds takes place in a competitive way, being based on an evaluation of applications from individual researchers, groups of researchers and institutes.

The Latvian Academy of Sciences is now an autonomous non-profit body financed by the state. It is a part of the national research system [4]. Its responsibilities are:

- to give advice to the government and parliament about research questions and to participate in the development of science policy in Latvia;
- to encourage studies in basic and applied sciences, placing particular importance on interdisciplinary research;
- to encourage the study of Latvian history, culture and language, as well as of the use of the natural resources of Latvia, possibilities for their optimal utilization and protection of the environment;
- to forecast the development of social processes in Latvia;
- to be an independent expert and adviser on principal problems of Latvia and the Baltic area as a whole;
- to publish scientific and scholarly works;
- to arrange scientific meetings and seminars,
- to favour and assist co-operation betwen Latvian and foreign scientific centres and function as the Latvian partner in international organizations, such as ICSU and Allea.

There is a special need for strengthening fields of research, that were not tolerated or were based on a state ideology in the former political system.

The three Baltic states are in other countries considered to be a unity similar to the well-known present unity of the Nordic countries (which, it should be remembered, through the centuries were in a more or less permanent state of mutual wars and only became a brotherhood during the last 150 years). No such brotherhood existed between the Baltic countries, which, it should be remembered, only have been independent states in the period between the two world wars and after 1991. They have each their own language, of which Latvian and Lithuanian are related, but different languages, while Estonian belongs to a different language group. The countries were all part of the Soviet Union and had Russian as the common language. There was, however, no strong bonds between them. The academies of sciences of the three countries have now realized that the new independent states have to co-operate in order to prepare and facilitate the wanted integration into Europe. They, therefore, meet regularly and have launched three large joint research programmes in co-operation with the academies of the Nordic countries. I shall briefly mention these three research programmes as an example of an initiative to be recommended to other regional groups of states.

- One research programme covers the humanities and social sciences, the research fields where the gap to the West is most significant. The reason for that is the just-mentioned former ideologisation of economics, social sciences, and parts of the humanities, leaving these fields to start almost from scratch after the political changes in the countries. These research areas are important for management and policy-making and thus fundamental ingredients in the state-building process. In the Baltic states, as in other countries in transition, there are national languages, but also, both now and in the past, sizeable communities of speakers of other languages. In these circumstances, studies relating to national identity assume a crucial role in the development of the countries, internally, in relation to their neighbours, and in relation to the outside world in general. Therefore, studies in language, ethnic relationships, national history, archaeology, folklore and other cultural traditions should be cultivated on the soundest scholarly basis possible in order to ensure a harmonious development of the countries and of their integration into the social, administrative, legal, scientific and general cultural systems of Europe.

- The second joint programme deals with the Baltic Sea, since the Baltic states embrace more than 1/5 of the area of the Baltic Sea, including the coastal area around the Bay of Riga with the outlet of large river systems and a heavy pollution. Research projects should investigate the exploitation of the resources of the Baltic Sea, pollution, restoration and conservation, by monitoring ecosystems and environmental conditions with advanced techniques, data analysis and modelling. The aim is to establish a sustainable management of the Baltic Sea.

- The third joint Baltic programme concerns energy research. All three Baltic states depend on the supply of energy resources from other countries, especially natural gas from the NIS region. Estonia exploits oil shale, a production that is accompanied by severe environmental degradation. Lithuania operates the Ignalina nuclear power station, the safety of which is questioned. Latvia is entirely dependent on import of energy and fuel. A stable energy supply is of great importance for these three, as for all other countries, combined with measures of conservation of energy and of reducing environmental side effects of energy production and transportation. This programme also has as a goal to prepare the integration into Europe, that is into European networks of energy supply.

I think that these initiatives of the academies of the Baltic countries deserve mentioning here as an example of regional co-operation in order to ensure a future stable development, internally in the countries and externally as part of the European community and its organizations such as the European Union.

A special problem is, how can, in the new reality, the high level research carried out at some of the former academy institutes be saved and further developed?

Wilhelm von Humboldt's classical pronouncement of the unity of research, teaching and study has been questioned, especially in countries where the system of higher education is under transformation into a mass-education system. In my view, integration of research institutes and universities and other institutions of higher education should, nevertheless, be encouraged wherever possible, since I see many advantages for both research and higher education in maintaining the unity of research and teaching. Besides, it is a way of getting the most out of scarce resources. There are, however, institutes that for one reason or another do not fit into the university structure, but nevertheless should be preserved, for national and sometimes also for international reasons. In the Baltic countries this is done by establishing centres of excellence, technological or innovative centres and science parks, a very difficult task in countries where research-based industries are practically non-existent. Here, assistance from the international community is needed.

The situation is complicated by the existence of a third kind of state research institutes, those of different ministries: industry, health, environment, etc. Many of these institutes were, in the former system, kept in total isolation from the academy and university institutes being occupied with areas that were considered to be of strategic importance. And they were, as one may say, over-populated in the same way as some of the academic institutes, not least because of enormous administrative staffs. These institutes are necessary for management of natural resources, health care, the social system, etc., but have to find their place in a new much contracted system.

I shall mention a new initiative from my own country that may serve as an example of how this problem may be solved. The Danish government has decided to establish a GEOCENTRE [5], that is to collect under one roof the Institutes of Geography and Geology and the Geological Museum of the University of Copenhagen (the Ministry of Education), the Geological Survey of Denmark and Greenland (the Ministry of Energy and the Environment), and a research institute (the Ministry of Research). This will create a Centre with a staff of about 600, which covers all fields of the geo-sciences needed in Denmark. The Centre wil have advanced laboratories, which are out of range for the single institution, and it will benefit from a tight co-operation with regard to library, general management, workshops, field equipment, etc. This is, on the one hand, an efficient use of public resources, on the other a favourable environment for interdisciplinary research in the geo-sciences.

I hope that this exposé of the situation in the Baltic states has elucidated a number of the problems facing the science systems and the academies in countries in transition. I have also endeavoured to show that academies of sciences, even if they are transformed from powerful centres of research to learned societies, still have a role to play. They can no longer base their activities on a strong economy and a large administrative staff. Their rate of success depends on the active involvement of the members. I am, however, afraid, that the members of the traditional types of learned societies have more than enough to do in their personal research, teaching and other professional functions and have little time to spare for the academies. This is the case in my own academy, where, with an administrative staff of only four persons, it is very difficult to involve larger groups of members, for instance, in the work of the international organizations. I am, however, impressed by the active involvement of the members of the academies of the Baltic states in the promotion of their research and its transfer to society and I hope that this will also be the situation in other countries in transition.

A precondition for the acceptance of the academies as advisers in questions of great national importance is, that there is in the scientific community and among the authorities no points of criticism concerning the status of their members, and that their autonomy is fully recognized. It is, therefore, important that the election of members takes place totally independent of authorities, institutions and organizations. Most of the academy members have positions in universities, research institutions and in industry and can thus, strictly speaking, not be considered to be 100 % independent. However, academies as bodies, and especially the traditional academies, which function as learned societies, are independent in a general way being self-elective and representing a continuity that is not found in research councils and other advisory bodies, where members are replaced at regular intervals. These are qualities that should be exploited, not least by countries in transition.

What is then the role of the academies of sciences in countries in transition ?

This can be said in the few words by Professor Jacquinot qouted at the beginning of my talk: *the academies should be the conscience of the scientific world or the scientific conscience of the world.* The academies should, based on the scientific capacity of their members and on their international contacts, find ways to alleviate the immediate problems and furthermore have the visions and prepare the plans for the future development of the research systems, which can be realized when conditions improve. I wish once more to refer to the Baltic states and to the co-operation established between the academies of sciences in these three countries. I think that this example of regional co-operation between academies should be followed by academies in other countries in transition, not only in the interest of the scientific community of the countries, but also as a way to build bridges between different cultures.

It is evident, that high priority should be given to safeguarding the research potential of countries in transition and to find solutions to current problems, such as the survival of the advanced research groups and institutes, the maintenance and, in many cases, the reconstruction of the infrastructure of the research system, the prevention of internal and external brain drain, etc. It is, however, also evident that these problems cannot be solved in the necessary pace in the contracted economy of these countries. Assistance from the international world is obviously required. It is definitely in the interest of the European countries and organizations, such as the European Union, to offer this assistance in order to facilitate a peaceful evolution and integration into the European system.

The evaluation of the Latvian research system organized by the Danish Reseach Councils, presented a number of recommendations to the international community. These, I think, are also valid in the case of other countries in transition and are therefore presented here as examples of possible international initiatives:

- There is a need for mobility schemes for junior as well as senior scientists enabling them to spend extended periods in other countries and for inviting scientists from other countries to work in the countries in transition. The sending of young scientists to other countries is one of the fastest and best investments in the future, but also holds the risk of inviting to a brain drain, when the conditions offered to the young researchers at home cannot compete with the conditions offered to them abroad.
- It is therefore important that there are grants to young scientists, which can cover the labour costs in the home institutions and make a return attractive.
- The international organizations should arrange workshops and summer schools in countries in transition with the purpose of updating skills and assisting in the transfer of methods.

- Funds should be made available for researchers from countries in transition to participate in international symposia, seminars, etc.
- The scientific community in countries with a contracting economy has great problems in maintaining its libraries. Help should therefore be offered to acquire the new scientific literature.
- Similarly, in these countries there is a lack of spare parts to scientific equipment and a lack of means to purchase modern instruments. Help should therefore be offered to update and renew instrumentation, computer networks and infrastructure in general.
- Help to participate in joint research projects, in part in the form of agreements with "twin" institutes in other countries, in part by participation in research programmes, as for instance the framework programmes of the European Union.
- An especially acute problem in a period of transition is the survival of top class scientists and institutes. They have much to offer to the international community. International assistance could take the form of help to introduce these institutes and their services to the international market, or of an invitation to take part in international programmes, such as the framework programmes of the EU, and of twin arrangements. The report presenting the Danish evaluation of the Latvian science system has the form of a catalogue, which, even though it is four years old, still may be of use in the search of a Latvian partner.
- Assistance in examining if there are patents, author's certificates, processes and products which could represent a value in the international market. International co-operation and support should be offered, wherever it is possible.

Finally, I find that the science academies of the world have a special role to play with regard to questions of global importance, such as the population problem, the sustainable exploitation of the Earth and its resources including the food supply to future generations of inhabitants of the Earth, the consequences of climatic changes, etc. The academies of sciences have an obligation to follow these questions closely, to supply governments and the public with the best, research-based information about these problems, and propose solutions, whenever it is possible. This should involve all academies, in regional co-operation, as for instance takes place between the Nordic countries, in broader scale co-operation, as is for instance by Allea, and on a more gobal scale, as it is represented by the InterAcademy Panel. The contributions by the academies of sciences to the United Nations Conference on Population and Development in Cairo in 1994 [6] and to the UN Habitat II Conference in Istanbul in 1996 [7] are examples of such activities. The academies of countries in transition should, even if they may feel that they have enough to do just to survive, join these activities of general global importance. I am happy to see, that the Statement by the World's Scientific Academies on Science and

Technology and the Future of Cities presented to the Habitat II conference, was signed by 72 academies from countries in transition, developing countries and the industrial countries. This, I feel, confirms that the academies of sciences accept that one of their responsibilities is to act as the scientific conscience of the world.

References

1. Evaluation of Estonian Research in the Natural Sciences. Stockholm 1992. Reports on other parts of the Estonian research system were published separately.
2. Latvian Research-An International Evaluation. *The Danish Research Councils*, Copenhagen, December 1992.
3. Evaluation of Research in Lithuania. *The Research Council of Norway*, Oslo, January 1996.
4. Kristapsons, J. & Tjunina, E. (1995) Changes in the Latvian Research System. *Science and Public Policy* **22.5**. Guildford, England
   Engelbrecht, J. (1995) L'Académie Estonienne des Sciences en transition. *UNESCO, Centre Européen pour l'Enseignement Supérieur* **xx**, 70-71.
   Millers, T. & Kristapsons, J., (1995) Création d'une Académie des Sciences de type classique en Lettonie. *UNESCO, Centre Européen pour l'Enseignement Supérieur* **xx**, no.4,72-83.
5. Betænkning om Geocenter. Betænkning nr. 1308. *Forskningsministeriet*, Copenhagen, December 1995 (English Summary).
6. Population Summit of the World's Scientific Academies. *The National Academy Press*, Washington, D.C. 1994.
   Population - the Complex Reality , a Report of the Population Summit of the World's Scientific Academies. Editor: Sir Francis Graham-Smith, F.R.S. *The Royal Society*. London 1994.
7. Science and Technology and the Future of Cities. A Statement by the World's Scientific Academies. *The National Academy Press*, Washington, D.C. June 1996.

INVENTIONS AS PROMOTERS OF SCIENTIFIC AND TECHNICAL PROGRESS

I. BOSTAN, V. DULGHERU
Technical University of Moldova
168 Stefan cel Mare Av., Chisinau, 2004, Republic of Moldova

1. General aspects

The development of society in the XX century brought creativity to the foreground of research. It happened because practically in all fields of activity the social demand for creativity became really urgent. In other words, re-evaluation of the place of creativity among modern values was done simultaneously with the recognition that creativity is the main resource for development. There are various types and forms of creativity, but technical creativity is considered to be unique in a way. Inventions, as a "personification" of technical creativity, with such parameters as novelty, value, applicability, are a driving force of technical progress. When the level of economic competition is high and its scope is wide, the price of new ideas correlated with the criteria of their efficiency, productivity, quality, etc., makes the orientation to novelty one of the main constituents of general evolution.

Let's imagine for a moment how our modern society would have looked if the great inventor hadn't realized the first fire through friction, if the greatest inventor of the world, Arhimed, with his outstanding invention, hadn't existed, if the Romanian architect Vitruvius hadn't invented the hydraulic turbine, Gittenberg - printing, in 1440, Denis Papin - the vaporous machine in 1695, Faradey - electric engine in 1822, Morse - the telegraph in 1843, the inventor nb. 1 of humanity Th. Edison - the incandescent lamp and phonograph in 1878, H.Coanda - reactive engine in 1910 and a lot of other pioneer inventions, which got some new directions off the ground in science and technology.

Contemporary society is realizing more and more fully that the sustainable living standard and good quality of life, to a great extent, depend on the efficient use of the creative capabilities of the members of that society. We are beginning to understand that, at the end of the XX century, the unlimited resources of human mind are becoming even more important than

Ch. Proukakis and N. Katsaros (eds.),
The New Role of the Academies of Sciences in the Balkan Countries, 171–182.
© 1997 *Kluwer Academic Publishers. Printed in the Netherlands.*

the depleting and exhaustible resources of nature. Thus, present-day conditions make an inventor a local figure of scientific and technical progress whose super-products, super-technologies are based on the latest inventions.

It is worth mentioning here that the unprecedented sophistication of technical systems and the growth in number of relevant problems to be solved are reasons for the intensification of creativity in engineering. The "gearing" of technical problems and engineering solutions is inter-dependent. The greater the range of technical problems, the wider the scope of engineering solutions, the latter in turn bring about new technical problems. It is quite to the point here to note the following: the number of technical systems doubles every ten years, the complexity of technical products doubles every fifteen years, the volume of scientific-technical information, which may be appropriate for inventions, doubles every eight years. But the period of elaboration of new products is reducing twice every twenty-five years. So, we can say that the volume of creative searching has been, and is, growing ten times every ten years, while the number of professional inventors is growing only three times in the same period, which means a shortage of creatively-oriented designers and constructors.

That is why it is evidently necessary to support technical creativity in every possible way; this type of creativity being a bedrock for many, if not for all, countries of the world. It is especially urgent for under-developed states and for economies in transition, an example of the latter being the Republic of Moldova. For instance, the diagram presented in figure 1 signalizes a total crisis in creativity, a field in which the Technical University

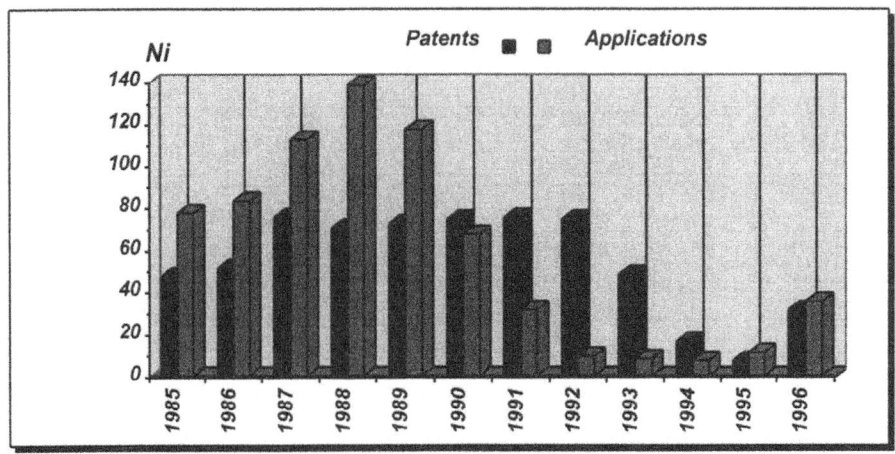

Figure 1. The dynamics of getting patents in the TUM.

of Moldova found itself in the last 6-7 years, being one of the most important sources of inventions from the Republic. For comparison, the number of applications for patents made by the inventors of the Technical University of Moldova in 1989 was 139, but in 1994 was reduced to 8. Probably it is explained more by the influence of economic factors rather than by that of creativity. A similar situation also exists in other scientific centres in Moldova and other countries of the former USSR.

The evolution of history shows that technical progress always was promoted, first of all, by scientists and engineers, by the people with a non-standard mode of thinking. It is owing to these people that new directions in science, new technologies, and new products appear. An accelerated pace of moral ageing of technical products keeps every inventor alert, in constant pursuit of novelty.

An invention is usually a new qualitative leap in the respective field. As a rule, a truly valuable and competitive invention is the "terminal link" in the chain of inventions. Experts say that a really valuable invention is one out of 50 others. For example, "Japan Research Development Corporation" analyzed 3,000 previously selected author's certificates registered during ten years. Then it picked out 300 of them, but only about 100 of those 300 led to super-new technologies, super-inventions. This is how things work in Japan, the country which is the first in the world in this field where about 200,000 inventions are made per year.

The complexity of problems which humanity faces today, need their solution in common, involving important intellectual forces from different countries. The solution of problems dealing with the exhaustion of mineral resources, energy crises and ecological catastrophes needs a stronger technical-scientific co-operation between various countries. The Scientific Committee OTAN could fulfill the function of co-ordinator of researchers from different states, favouring the organization of creative mixed collectives.

2. Concretion aspect

Everything mentioned above is shared by the authors of the given report and may be confirmed by their own long-term practical experience in making inventions. A thorough analysis of the state-of-art in mechanical drive in the late 1980s resulted in the new conception of the planetary precessional drive with the sphere-spatial movement of the pinion with multiple gear [1,2,3,4,5]. It was considered to be a totally original direction which shortly became developed and materialized. As a result, new inventions came to the world, not of extreme importance, but they promoted other, more than 130, inventions, among them 5 or 6 being the real winners in the race. The novelty of those inventions with the elements of know-how made it necessary to solve a set of problems related to the estimation of mechanical drive, such as:

- the development of a fundamental theory of a multiple precessional gear (about 100% of pairs of teeth may be in operation simultaneously);
- the bringing-together of a wide range of precessional drives that can be used in various technical areas;
- the development of high technologies for manufacturing gears, and of new methods for their quality control;
- the development of the foundations for calculating and designing precessional drives.

At the same time it became necessary to abandon some principles used in gears. For instance, in the elaboration of the manufacturing technology, the principles of geometrical theory of involute gear offered by Thomas Olivie were not appropriate. Those principles did not meet the requirements for the precessional drive, therefore, other principles were established that fitted the conditions for the sphere-spatial pinion movement in precessional drive. The priority fields for the application of the precessional drive are: working out technologies for extracting Ferro-manganese concretions from the world ocean bottom, cosmic space exploration, manufacturing of industrial robots, applications in mining industry and in various technical equipment for different purposes, etc.

2.1. THE ELABORATION OF ROBOT COMPLEX DRIVE MECHANISMS FOR THE EXTRACTION OF FERRO-MANGANESE CONCRETIONS FROM THE WORLD OCEAN BOTTOM

The high rates of development of the various sectors of the national economy lead to a rapid exhaustion of raw materials; the overall exhaustion of some of these materials is expected during the first decades of the third millennium. Partially, the way out of this situation can be ensured by discovering and utilizing new raw materials on a large scale. From this point of view, the world ocean can be considered an important inexhaustible source of raw materials which takes up more than 2/3 of the planet surface and the utilization of its assets is reduced.

A special interest presents the concretions of pure metals in the form of black tubers discovered in 1872 by the expedition carried out on the Queen Victoria's ship" Challenger"[6]. Similar concretions were discovered by C. Hun in the Atlantic Ocean in 1898 within the expedition on the first German research ship. These concretions in the form of tubers have a diameter 1-12 cm (figure 2b). In cross section these concretions have a concentric stratified structure. It proves that these tubers appeared in the process of deposition of each new layer around the centre of crystallization and it was a very long process. The manganese concretions typical for the world ocean contain on the average 27% Mn, 8% Fe, 1,4% Ni, 1,3% Cu, 0,2% Co. To be more precise: the deposits of raw materials at a depth of about 5,000 m make 15-75 kg/m^2 (figure 2b).

At present, the scientists work intensively on the technological and economical aspects of extracting, transporting and enriching the concretions. From 1984 till the disintegration of the USSR, a research team from the Technical University of Moldova (TUM) under the leadership of Prof. Ion Bostan, Academician of the Moldovan Academy of Sciences, Dr. habilitat., participated in the investigations within an all - union research programme concerning the elaboration of the technological devices for Ferro-manganese concretion extraction. That programme was set up by the Council of Ministers of the former USSR from 29.08.87 "Concerning the measures of ensuring the elaboration of the technological devices for the extraction of Ferro-manganese concretions from the bottom of the sea sector belonging to the USSR" N1007 - 236.

The objectives of collaboration of the research team of the Technical University of Moldova within this programme were the elaboration of a whole series of precessional reducers necessary for the creation of the robot complex for exploring and extracting Ferro-manganese concretions from the World Ocean bottom (figure 2a). The functioning conditions which are extremely difficult (the necessity to fill in the reducers with oil to compensate the ambient pressure of 50...60 MPa, a considerable growth of oil kinematics viscosity due to low t=2...4°C, slow speed of working elements which require big gear ratio U=100...600) practically make impossible the utilization of well-known mechanical transmissions. From this point of view, the precessional reducers elaborated by the research team from TUM possess a wide range of advantages:

⇒ the peculiarities of sphere-spatial motion of the pinion ensures minimum hydraulic losses and high efficiency;
⇒ the increased bearing capacity and kinematics accuracy due to the gear multiplicity (about 100% couples of teeth are engaged simultaneously);
⇒ the possibility to realize a big reduction ratio (U = 8...3600) in compact constructions;
⇒ the precessional gear with tooth-roller contact permits the elaboration of reducers ecologically pure in exploitation, i.e. essentially for the elaboration of concretions exploring technologies with a high level of environmental protection.

176

1 - 3 cm   3 - 6 cm   6 - 12 cm

Figure 2. The robot complex for exploring and extracting Ferro-manganese concretions from the World Ocean bottom.

Know-how in the elaboration of the multicouple precessional gear, manufacturing technology and control methods, and a large range of precessional transmissions diagrams belong to a research team from the Technical University of Moldova. During the last 15 years, the team patented more than 130 inventions.

The elaboration of working machines driving mechanisms is based on the diagram of precessional transmissions, presented in figure 3 [5,7] . The rotating motion of the crank shaft 1 is transformed into sphere-spatial motion of the block pinion 2 with two teethed crowns 6 and 7, which are rolling without sliding on the immovable and driven toothed wheel teeth 3 and 4. The teeth of crowns 6 and 7 are manufactured in the shape of conical rollers installed on axis having the possibility to rotate round them, and the teeth of central wheel 3 and 4 have non-standard convex-concave profile (figure 3,b,c). The new method of processing the teeth by grinding and milling guarantees the ideal accuracy of the profile and teeth pitch. The disassembled reducer is shown in figure 4.

The authors have elaborated the technical documentation of 9 precessional reducers for various drive mechanisms of the extracting robot complex. From the elaborated range of precessional reducers we have implemented the "zero" series of 3 type-dimensions and performed the experimental research which showed high performances.

2.2. THE ELABORATION OF THE AEOLIAN UNITS FOR WIND ENERGY RECOVERY

Modern energy is especially based on such fossil fuel as coal, peat, oil ,gas and uranium. The share of fossil fuel-coal, oil and natural gas - is 95%, of water 4%, of uranium 1% in the overall energy consumption [6]. At present, man exploits power, natural resources which are almost exhausted or can be utilized only in reduced quantities (hydraulic energy).

Therefore, mankind is in the process of seeking new forms of energy, the so-called non-traditional sources of power - solar, aeolian, hydraulic, etc. These renewable energy resources are practically inexhaustible. It is necessary only to elaborate high efficiency technologies of their recovery and utilization. The utilization of renewable energy is opportune from two points of view: the economic aspect in the first place because the exhaustion of traditional energy sources forces us to seek new sources; not less important is the ecological aspect.

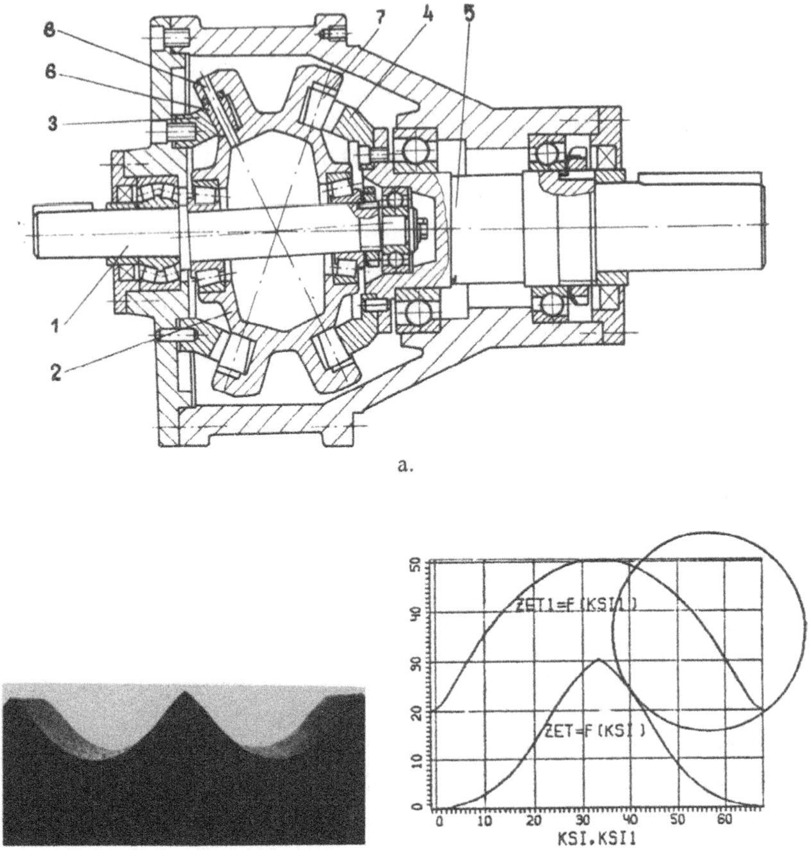

a.

b.

c.

Figure 3. Precession reducer

a. 1 - single throw crankshaft; 2 - planet rolling unit of pinions; 3 - fixed gear wheel; 4 - movable gear wheel;
5 - driven shaft; 6,7 - rolling rims; 8 -a xes

b. teeth of crowns 6 and 7 in the shape of conical rollers

c. teeth of central wheel 3 and 4 have non-standard convex-concave profile

Figure 4. Precessional reductor taken apart

One of the main sources of renewable energy is wind energy. In conformity with the estimations, the world reserves of aeolian energy are $18*10^{12}$ kW [6]. The advantages of aeolian energy in comparison with other renewable energies are: relatively reduced cost price, possibility to use stations in regions difficult to access, comfort in maintenance and autonomous functioning. For the last 15 years, the research team from the Technical University of Moldova under the leadership of *Prof. Ion Bostan*, Dr.Sc., Academician of the Moldovan Academy of Sciences, has carried out ample research to elaborate the units for wind energy recovery (figure 5a).

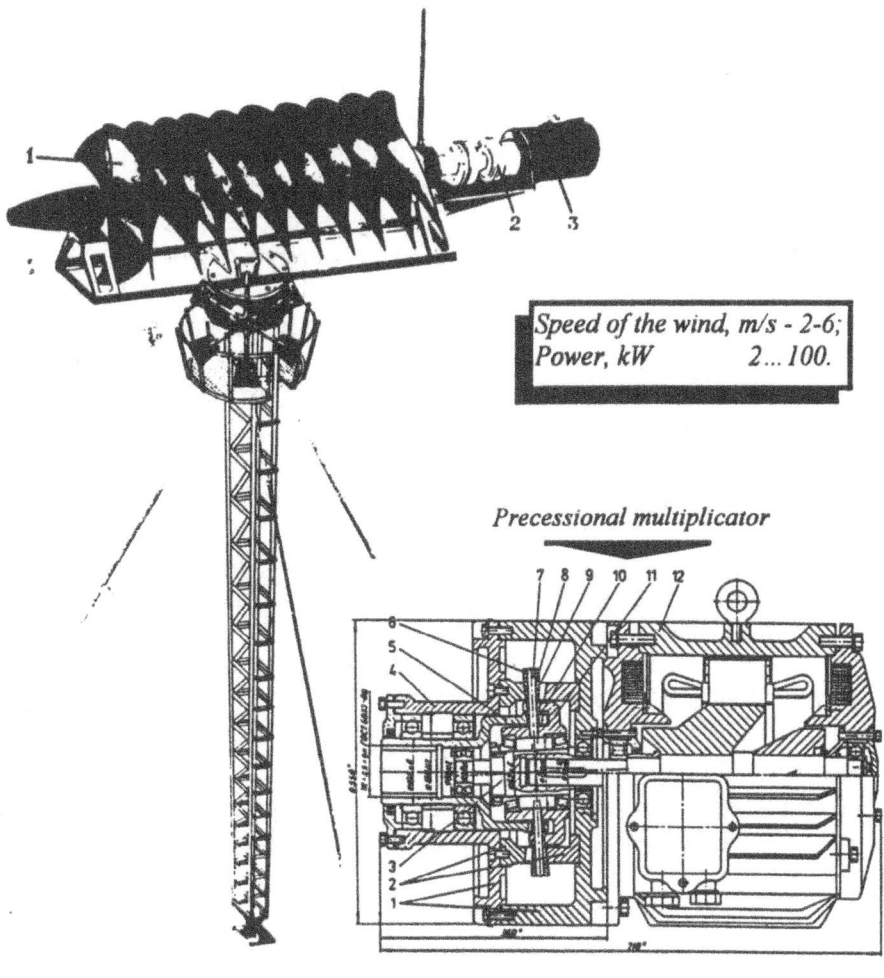

Speed of the wind, m/s - 2-6;
Power, kW   2...100.

Precessional multiplicator

a.   b.

Figure 5. Electrowind power stations.
1. Turbine; 2. Precessional multiplicator; 3. Electric generator.

The multiplicator is a very important unit of the electro-aeolian set. It is destined for the rotation multiplication of the working organ up to the working frequency of the electrical generation. The multiplier must follow a series of conditions such as: reduced dimensions and mass, high efficiency, simple technology of implementation and assembling, efficient work in extreme conditions - high or low temperatures, absence of lubricant, etc.

The investigation in the field of planetary precessional transmissions allowed the elaboration of the precessional multiplicator (figure 5b) for the electro-aeolian set whose power is **P=8 kW** and **P=16 kW** respectively [1,8]. As a basis for the elaborated multiplier, the precessional transmission scheme with the plat pinion has been taken. This pinion gears from both sides with two fixed rack wheels, their number of teeth being equal. This permitted the increase of the bearing capacity as a result of the gear multiplicity growth (about **100%** of teeth couples are simultaneously gearing).

The working organ is another unit of the aeolian set. The existing working organs (in the form of propeller, rotate cylinder or merry-go-round, etc.) do not ensure high efficiency utilization of the wind and efficient functioning at small speeds of the wind. The research in this field allowed the elaboration of a helicoid rotor with special profile 1(figure 5a) of the sails which ensures efficient functioning at small speeds (beginning with 2 m/s) developing large power due to the injection effect of the lateral air masses. This would present interest for European countries, where the speed of the wind is lower than 2...6 m/s.

All problems connected with the elaboration of the precessional multiplier have been solved at the Technical University of Moldova:
- *"gear synthesis - profil study - fabrication - control";*
- *determination of the working surface and profile geometry in a cross-section of the helicoid rotor, which ensures maximum efficiency of the wind kinetic energy utilization.*

2.3. PERSPECTIVE ELABORATION

Present and future overproduction performs great harm to the environment. Therefore, we must assume as a basis the ecological factor of all the elaborated technologies. Some perspective elaboration in this direction has been worked out:
- precessional drive devices with teeth gear and radial studding bearings for the extracting robot complex to be used on the bottom of the world ocean, which function utilizing as lubricant sea water substituting oils that pollute the ocean;
- the world energy crisis makes new demands for energy producers.

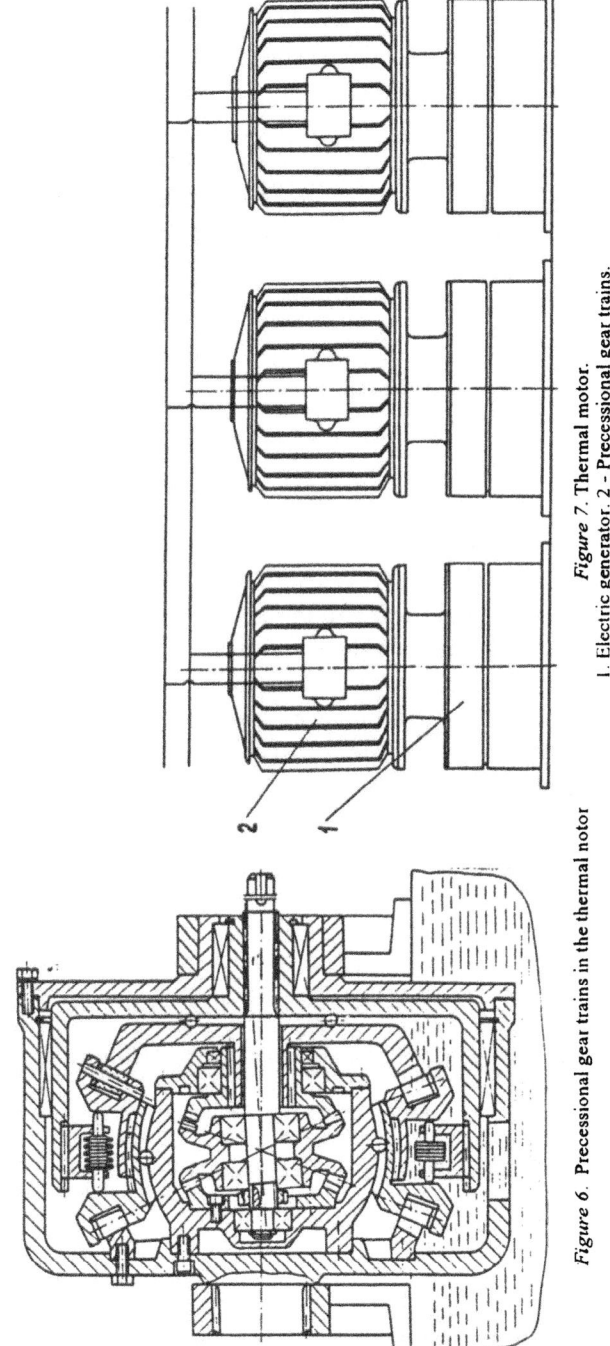

Figure 7. Thermal motor.
1. Electric generator. 2 - Precessional gear trains.

Figure 6. Precessional gear trains in the thermal notor

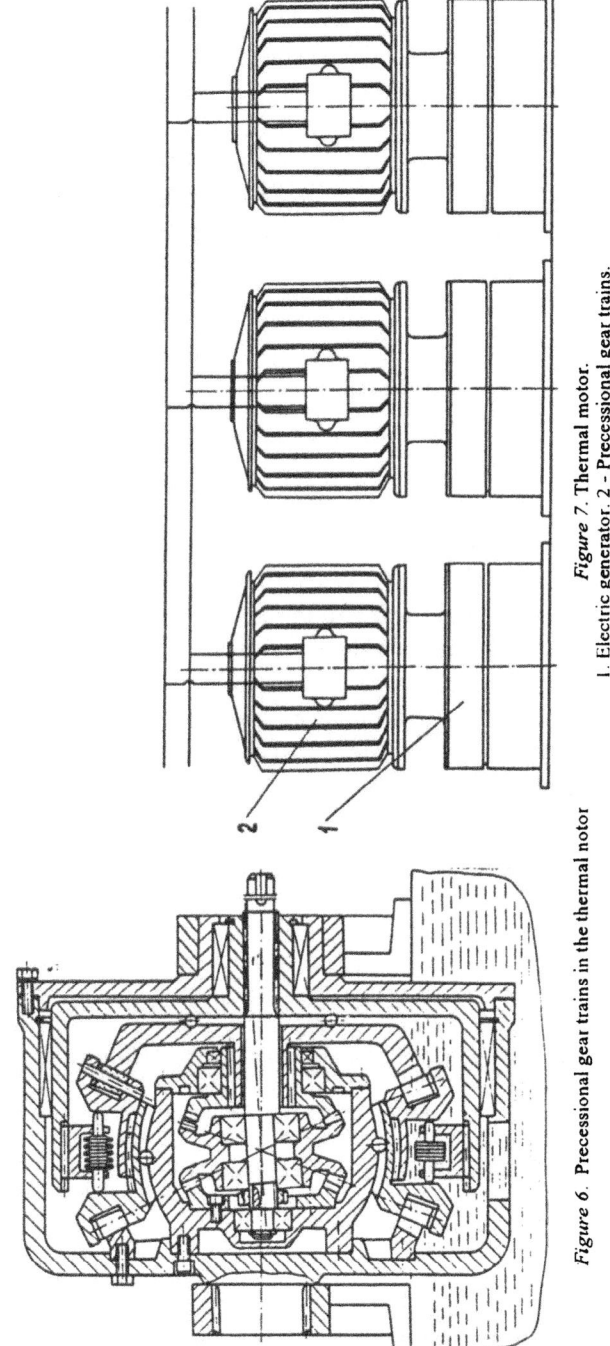

181

182

It is very important to use in this direction technologies based on the utilization of thermal waters energy. A special device [9] was elaborated to transform thermal energy into mechanic energy based on precessional gear with toothing rim of the planetary pinion executed, for example, in the form of a silphon filled in with a working agent or some component elements performed from metal which possess the memory of the shape (nickel-titanium alloy) (figure 6,7). The rotating planetary pinion brings the working elements into various media with a difference of temperatures. The energy produced by these elements is transmitted to the working device.

As a result, the efficiency of our participation in the international economic circuit depends on the position in the market of patents. As proof that the Republic of Moldova has enough technical potential, at the last 3 World Exhibitions on Innovations and Modern Technologies in Belgium, Brussels, and Pittsburgh, USA, and International Exhibitions on Innovations and Modern Technologies in Iassy (Romania), and Budapest (Hungary), Moldovan inventors were awarded over 30 gold medals and special prizes, that make up 80% of the presented elaboration.

The research team from TUM solved practically the range of problems related to the "gear synthesis - profile study- fabrication - control". All patents ensure the know-how regarding the elaboration of precessional gear, the synthesis of kinematics schemes, the elaboration of an efficient fabrication technology and monitoring of the production quality. The UTM Collective wants to collaborate in mixed collectives for the solution of urgent problems mentioned above.

References

1. Bostan I. (1992) *Precessional epicyclic gear with multiple engaged tooth couples.* Stiinta, Chisinau
2. Bostan I., Glusco C., Dulgheru V. and Oprea A. (1987) *Precessional epicyclic gear trains,* Stiinta, Chisinau
3. Bostan I. (1988) *Precessional gear engagement,* Stiinta, Chisinau
4. Bostan I. (1989) *Precessional gear - engagement,* Patent RU 1563319 (patent MD 540).
5. Bostan I. and Dulgheru V. (1992) *Device for wheel control,* Patent RU 1732138 (patent MD 631).
6. Felix R.Paturi. (1975) *Baumeister Unserer Zukunft: Kuhne Projekte der Forscher, Erfinder und Ingenieure in aller Welt,* Econ Verlag, Dusseldorf-Wien.
7. Bostan I. and Dulgheru V.(1989) *Planetary Precessional gear.* Patent SU 1714249 (patent MD 636).
8. Bostan I. and Dulgheru V.(1992) *Electrowind plant,* Patent SU 1760151 (patent MD 632).
9. Bostan I., Dulgheru V. (1991) *Appliance for conversion thermal energy into mechanical energy,* Patent SU 1671956 (patent MD 627).

SOME ASPECTS OF REFORMATION OF THE NATIONAL ACADEMY OF SCIENCES OF UKRAINE

S. A. ANDRONATI
A.V.Bogatsky Physico-Chemical Institute of the National Academy of Sciences of Ukraine
86 Lustdorfskaya doroga, 270080 Odessa, Ukraine

Ukraine was one of the regions of the former Soviet Union. The system of the republic organization of the investigations and elaborations was the part of the All-Union system. Ukraine became an independent State in 1991. It inherited the scientific system that had a number of defects and did not correspond to the requirements of the country. These defects included various deformations concerning development of the investigations in the field of the natural sciences, humanitarian and especially applied sciences. For example, the share of work directed to the defence of the country was too large. Social-humanitarian investigations were extremely politicized and developed in isolation from world science. The juridical defence of the intellectual property had not enough guarantee.

Scientific schools in the field of mathematics, mechanics, physics, chemistry, biology, technical sciences have gained recognition all over the world. Results of the fundamental investigations and elaborations of the Ukrainian scientists in the field of the materials science, weld and cutting of metals are the base for the creation of unique technologies and materials. Many of these are applied in the different countries of the world.

Researches in the field of biochemistry and physiology of the nervous system, in the series fields of medicinal science, bioorganic and supramolecular chemistry keep the advanced positions in the world science. Just as a level of many investigations and elaborations is lower than the world level.

During the last years, the state financial resources for science have sharply decreased due to the economic crisis. Assignations in 1995 were less than 1% of the gross domestic product as compared with 3.1% in 1990.

Actually, the financial resources for investigations and elaborations according to the contracts between scientific organizations and enterprises of different industrial branches have been exhausted due to the crucial state of industry.

Ch. Proukakis and N. Katsaros (eds.),
The New Role of the Academies of Sciences in the Balkan Countries, 183–191.
© 1997 *Kluwer Academic Publishers. Printed in the Netherlands.*

During the periods of industrial depression in the developed countries, the share of state investments in science increased. We have no such possibility in Ukraine.

As a result of financial reduction, the number of employees in the scientific sphere has decreased, and many of the perspective researchers have had to leave Ukraine for work abroad. Thus, during 1991-1994, the quantity of scientists in the Odessa region had decreased by 32.2% and the quantity of Ph. doctors had been reduced by 19.5%.

Equipment and devices are out of date, resources of reagents as well as materials for scientific research and elaborations have exhausted, the pilot-experimental base is wrecked, the level and efficacy of researches and elaborations is decreasing. The moral costs are not accounted for.

The scientists are the elite (although not highly payed) in the civilized society, but the profession of the scientist in Ukraine as well as in other states of the former Soviet Union has lost its prestige. The influx of young people to the post graduate studentship and to the large number of the faculties of the higher educational institutions is decreasing. One of the most anxious manifestations of the science crisis is the closure of activity of some scientific schools. The analysis of the state of scientific and technical potential of the South region of Ukraine testifies that the series of scientific schools that were recently famous all over the world have stopped their activity drastically. Thus, for example, the world famous mathematical school of Professor M. G. Krein has been lost. Prof. M. G. Krein died and a large part of his gifted pupils are working abroad.

It should be unreal to suppose that in the nearest future the possibility will be revealed for the state to pay out finance as much as it was in 1990.

Nearly 50% of the national scientific expenses of USA are from industrial corporations. It is an extremely optimistic point of view to expect serious investments in science from Ukrainian businessmen. Taking into account the humanitarian importance of the assistance for Ukrainian researches from the international scientific foundations, it should be noted that the share of these grants in the total financing for the scientific investigations does not exceed 1%.

Is it possible to maintain science in Ukraine on an adequate level in such a crucial financial state?

The main directions on the reformation of the investigation and elaboration organizing system with the purpose to increase its efficacy have been discussed during the all-Ukrainian meeting on the problems of science development (Kyiv, February 1996).

With the purpose to overcome crisis in Ukrainian science and to intensify its role in structural reconstruction of state economy, the following is envisaged:

- to develop the idea on the development of socio-humanitarian sciences;

- to develop the conception on the creation of the system on scientific-technical information;
- to develop the conception of scientific and scientific-technical politics;
- to bring the structure of the state scientific-technical potential in correspondence with state priorities;
- to intensify the role and responsibility of the National Academy of Sciences of Ukraine (NASU) in the field of fundamental investigations and the role of the Ministry of Science and Technologies Affairs in the field of the development and the optimal use of the scientific and technical potential;
- to develop the principles for the financing of science and innovation politics for training of scientific specialists and social protection of the researchers.

In the former Soviet Union, the lion share of scientific and experimental and design work was carried out by the Ukrainian scientists in favour of the military-industrial complexes. Now it is absolutely unwarranted to support such a level of "militarization" of science. The scientific problems should be reconsidered and priorities should be determined. The limited financial and material resources should be concentrated on these priorities. The concentration of scientific and technical potential on the decision of priority problems could be reached with financial and administrative mechanisms, with the complexity of activity of academic scientific institutions, their pilot-experimental base, high education institutions, branch scientific-research institutions and design bureau industries.

The system of research and elaboration organization in Ukraine changes drastically. The novel system of science organization consists of:

- The Council on the problems of science and scientific-technical politics at the President of Ukraine. It is headed by the President of Ukraine;
- the Ministry of Science and Technology Affairs and subordinated regional centres of scientific-technical information. The Ministry of Science and Technologies Affairs in the regions at the regional administrations;
- the state foundation of the fundamental investigations at the Ministry of Science and Technologies Affairs;
- the National Academy of Sciences of Ukraine, as the state institution, is responsible, first of all, for the development of fundamental researches, its regional scientific centres;
- the Ukrainian Academy of Agricultural Sciences;
- the Academy of the Medicinal Sciences of Ukraine;
- the Academy of the Pedagogical Sciences of Ukraine;
- the Academy of the Juridical Sciences of Ukraine;
- the State Innovation Foundation and its regional divisions;

- the Ministries and Departments of Ukraine (State Standard, State Patent, Higher attestation commission etc.).

Science organization is realized at three levels: state, branch and regional.

Main means for science control are the financial key factors - financing of national, state, field and other programmes, grants.

The National Academy of Sciences of Ukraine carries on the wide range of work on the reorganization of Academic science. The priorities in the research directions have been revised. The general directions of the activity of the Academy institutions are those where investigations are carried out on a world level, where results are significant for world fundamental science. Important consideration is given to the directed fundamental researches with results for creation of high technology and perspective materials and in the nearest future they will favour the progress of state economics as well as the decision of tasks of stable development of society. The priority investigations and elaborations of the Academy institutions are those which are directed to problems of energetics, economy of resources, problems of agricultural complex, materials science, protection of human health.

Research work with its significance for the cultural revival and for national progress, for the building of the independent state, is of paramount importance in the field of the socio-humanitarian sciences.

The availability of the powerful pilot-experimental and in many cases the industrial base in the structure of institutes was the important difference between the NASU and Academy of Sciences of the USSR and Academies of Sciences of other republics of the former USSR. Such a structure of institutes allows the realization of a complex of investigations from fundamental research to applied elaborations and to pilot-industrial work and then to the organization of production.

There are a lot of problems and difficulties now for the activity of the pilot and design base that deals with the general economic situation in the State. The important task for the Academy is to maintain and develop its pilot base at the novel economic conditions. One of the possible ways to solve this problem is private and foreign investments e.g. the assistance for activity of design offices, pilot plants, engineering centres based on their joint scientific-production activity.

The activity of the E. O. Paton Institute of electric welding (Kyiv) in the above-mentioned direction is the most impressive. There are many joint ventures created by this Institute in Ukraine as well as abroad.

The A.V. Bogatsky Physico-Chemical Institute (Odessa) includes in its structure the Ukrainian-Belgian joint venture "InterChem" that realizes development, industrial output and realization products of fine chemistry (reagents for scientific investigations, medicines, catalysts etc.) in domestic as well as European markets. The mutual activity is profitable, it ensures the normal functioning for the pilot-experimental base and allows for the realization of significant engineering re-equipment and extra financing for elaborations of the Institute.

Fundamental research, elaborations and production output by the created technologies are realized by scientific concerns of the National Academy of Sciences of Ukraine based on the Institute of super-hard materials and the Institute of monocrystals.

The important role in the reorganization of science of the Ukraine Academy is given to the relations of consolidation between institutions of the NASU and branch institutes and enterprises. These relations are based on the contracts about the mutual activity of the Academy and corresponding ministries and departments. In the series of cases, institutes of the Academy or branch institutes become institutes under the double subordination. Herewith, the Academy realizes scientific-methodical ensuring and the Ministry or Department provides the material-technical ensuring of such an institute. These relations allow for the more effective use of funds for the development of science and technology. Moreover, the above-mentioned relationships make investigations and elaborations more actual for the Ukraine economics.

The main part of higher qualification researchers, above 3/4 of professors and approximately 60% of Ph.Dr. in our region, are employees of higher educational institutions. The significant scientific personnel potential could appear as serious stimuli in the increase of the efficacy of researches as well as elaborations.

However, scientific groups of higher educational institutions are in a crucial economic situation now. First of all, this fact is due to the absence of financing for scientific work and to the extreme dissipation of resources. In many cases, the inter-chair and inter-institute co-operation for the decision of scientific tasks is absent.

The concrete positive role in the consolidations of efforts for the decision of important regional problems is played by the sections of the South Scientific Centre of the Academy. They develop programs on such problems and provide an interdisciplinary, complex approach to the decision of tasks.

With the purpose to increase efficacy of fundamental investigations and elaborations in the field of natural and social-humanitarian sciences, the interaction between Institutes of the Academy of Sciences and higher educational institutions of Ukraine (universities first of all) take on a special significance. Profits from such an interaction are especially evident in the modern conditions. This elaboration supposes that Academy scientists will participate in the training of personnel with higher education and specialists with higher professional skill for work in the different fields of national economy, science and education. Professors and teachers of higher educational institutions will get an opportunity to work at the best laboratories of the Academy and to use the equipment and devices of scientific institutes which, as a rule, are well equipped compared with chairs and laboratories of the higher educational institutions. Through the collaboration of universities and scientific institutions of the Academy, the level of personnel training as well as the level of researches will be higher. As for the applied elaborations, the organization of the realization of results of works in practice will be simplified. It should be noted that such combined efforts of the Academy scientists as well as scientists of higher educational institutions allow for the concentration of the scientific and technical potential for the decision of the most important and urgent tasks.

The organization forms of such joining differ: institutes under the double subordination (Academy and Ministries of education), joint chairs, branches of chairs, Academy- University complexes etc.

For example, the A. V. Bogatsky Physico-Chemical Institute of NASU has a long-term relation with the Odessa I. I. Mechnikov State University. The Academy-University scientific complex on the problems of organic and bioorganic chemistry has been acting for over 20 years. It consists of six scientific departments and one pilot-technological department of the Institute, Chair of organic chemistry and Problem laboratory of the University. Activity of this complex has given the significant fundamental results in the field of organic, bioorganic and supramolecular chemistry and on their base the creation of a series of effective medicals and a large number of reagents for scientific purposes. This is the production of the joint venture "InterChem". During the last years, 13 Doctors (Professors), 105 Ph.Dr., over 300 specialists with higher education have been trained within the Complex. The Institute and the University through the expansion of co-operation have created in the Institute the Branch of the University Organic Chemistry Chair and since 1995 the training of specialists on the "pharmaceutical chemistry" speciality has started.

The commercial structure "InterChem" is included in the given Complex. It allows to successfully decide problems of material and engineering provision for the scientific activity and for the training of specialists.

It is very important that this system allows for the results of investigations and elaborations created by employees of the Complex to be applied in industry more quickly. The commercial enterprise has the following evident profits:

- the availability of intellectual property created within the scientific departments of the Complex;
- the highly skilled provision for technology processes;

- the possibility to predict the perspectives on the development of the corresponding branch of fine chemistry.

There are a lot of other examples in the Odessa, Nikolaev and Kherson regions on the traditional relationships and joint activity of scientific institutions, of higher educational institutions, and enterprises in the field of heating physics and heating engineering, refrigeration engineering, biology and medicine, agricultural sciences etc. These relations are urgent due to the reorganization of the scientific and engineering activity in Ukraine.

The important task of the regional organization of the mentioned activity is to revive and develop these relationships, and to adapt them to the novel economic situation.

The organization of investigations and elaborations for the purposes of economic and social development of the corresponding regions is the main task for the scientific and engineering development.

The efficacy and necessity of such a system has been confirmed by the long-term experience accumulated by the NASU.

For 25 years, the regional scientific centres (North-Eastern, Western, South, Donetsky, Kyivsky, Pridneprovsky) were organizing the complex investigations and elaborations directed to the decision of social and economic problems of their regions.

One of the most important tasks for the NASU is the performance of the experts decision on the different problems of economics and social development, and on the perspectives of science development. The functions of the expert commissions are important in the activity of the regional scientific centres of the Academy. The examples could be given according to the activity of the South Scientific Centre (SSC) of the NASU.

This Centre was founded in 1971 and it has consolidated efforts of scientists and specialists from South Ukraine to develop the actual regional problems on the preservation and rational use of resources of the Black Sea, Dniestr, Dnepro-Bugsky leman, of the large irrigation systems, of the power supply system etc.

In some cases, the decisions of expert commissions did not correspond to the decisions thrusted by Moscow or Kyiv concerning the series of the large economic and nature-transforming projects. Such facts took place, for example, in the case of the project on the creation of the Danube-Dnepr irrigation system, on the creation of the nuclear heat and power plant near Odessa and on the creation of the Berezovsky fertilizers plant. The scientifically grounded conclusions promoted the formation of an adequate public opinion concerning this kind of projects non-justified economically and ecologically.

The formation of the draft state programme on the social and economic transformations of South Ukraine has broadly been promoted. This work should favour the directed structure reorganization of scientific as well as scientific-technical institutions of South Ukraine.

The reorganization should be directed towards the creation of the integral national economic and scientific-technical complex of the region. This Complex will favour the solution of the most part of the socio-economic problems of regional life.

The efficacy of the regional management system of scientific and technical progress depends on the possibilities of the administrative methods of the economic and social sphere management.

It is evident that these management methods for the scientific and technical progress are impossible now. The main emphasis in organizing regional scientific and technical activity should be on the economic methods.

I should like to go into the details of the regional level of the management of science and technology in the present situation.

General management at this level is realized through state regional administrations. The Committee on the problems of science, technology and industrial politics subordinates both the regional state authorities and the Ministry of Science and Technology Affairs,and has been set up in Odessa. The Committee, the South Scientific Centre of the NASU and the Odessa Centre of scientific and technical information subordinates the Ministry of Science and Technology Affairs and form the system realizing the State scientific and technical politics in the region.

Scientific organizing activity will be realized according to the conception of scientific and technical politics in the region. This conception is developed now and it includes the following statements:

- on the priority directions of fundamental and applied investigations including requirements of regional economics;
- on the main scientific schools and perspectives of their development;
- on the procedure of financing of regional investigations, scientific and technical programmes, some researches and elaborations;
- on the intensification of the scientists' and specialists' integration in the scientific institutions and design bureaus of State Academies of Sciences, higher educational institutions, branch scientific-research centres, different scientific and technical associations
- on the system realizing scientific and technical advances in the regional national economics and innovation politics;
- according to the priority directions of the science and technology development to form the complexes of the related scientific institutions, scientific-production and scientific-educational associations, design bureaus and manufactures irrespective of the departmental affiliation. These complexes should be ensured the whole cycle of fundamental researches and applied works, pilot-experimental and organizational-

technical arrangements for the realization of elaborations in national economics;
- on the arrangements to increase the role of the socio-humanitarian sciences for the solution of problems concerning creation of the independent State.

The local budget, and partially the state budget and other resources will be financing the base for the realization of the regional scientific and technical politics.

Regional departments of the state innovation foundation will finance the introduction of progressive technologies.

It is impossible to settle the important regional problems without the widespread international scientific and technical co-operation. For example:
- problems of preservation and rational use of the resources of the Black Sea and Danube, Dniestr etc. Work in this direction has just started.

Scientists and specialists from Ukraine and Moldova co-operate according to the Agreement between the Ukraine and Moldova Academies of Sciences. An intergovernmental Ukrainian-Moldavian working group on the problems of Dniestr basin has been formed.

The creative relationship between scientists of Ukraine and Poland has become stronger.

The maintenance of the positive elements in organizing Ukrainian science, its adoption for the market economics as well as for the real conditions and requirements, the building of an independent State - this is the sense of the reform.

TRANSFORMATION OF RESEARCH IN THE CZECH REPUBLIC:

The Role of the Academy of Sciences

VACLAV PACES
Academy of Sciences of the Czech Republic
Národní 3, CZ-11142 Prague 1, Czech Republic

1. Introduction

The Academy of Sciences of the Czech Republic was established by law in 1992 as the Czech successor of the former Czechoslovak Academy of Sciences. It is structured as a network of 59 research institutes and five supporting units and is staffed by 6,500 employees, approximately one-half of which is university-trained scientists and Ph.D. researchers.

The chief objective of the Academy is to carry out fundamental and strategic applied research in a broad spectrum of natural, technical, and social sciences as well as the humanities. This research is distinguished by adherence to high scientific standards whether interdisciplinary in nature or highly specialized. The Academy's institutes are also involved in education, primarily by supervising Ph.D. theses but also by providing post-graduate courses to young researchers and by teaching at universities. The Academy promotes contacts with both the applied research and industrial sectors in order to foster technology transfer and exploitation of scientific knowledge. Many joint international projects and staff exchange programmes with foreign partner institutions ensure the participation of Czech scientists in international activities.

After the political changes in 1989, the Academy was the first scientific and research institution in the Czech Republic to separate its decision-making, control and executive functions and introduce an evaluation procedure. The main decision-making body is the Academy Assembly; two-thirds of its members are representatives of the institutes and the rest are university scholars, government officials and other distinguished public personalities. The executive body is the Academy Council, chaired by the President of the Academy who bears ultimate responsibility for carrying out the Academy's mission. The Council for Sciences is mainly responsible for shaping science policy to be pursued by the Academy. Members of these

Ch. Proukakis and N. Katsaros (eds.),
The New Role of the Academies of Sciences in the Balkan Countries, 193–199.
© 1997 Kluwer Academic Publishers. Printed in the Netherlands.

bodies are elected, usually for four-year terms of office. For the evaluation of individual institutes, the <u>Academy Appraisal Commission</u> appoints independent, ad hoc <u>evaluation committees</u> consisting of prominent scientists from non-Academy institutes and universities, half of them foreign.

The Academy of Sciences is financed primarily from the state budget, which is approved by Parliament; in 1996, this amount was Kc 1,643 million. This sum is divided among the institutes according to the evaluation of their scientific results. Further sources of income include funds acquired from grant agencies in competitions of scientific projects. The Academy has also its own grant agency to support small-scale projects and projects of young scientists; it is open to applications from non-Academy scientists. The individual institutes can also acquire other financial resources by means of their economic activities.

2. History

The present-day Academy of Sciences of the Czech Republic builds not only upon the traditions of the former Czechoslovak Academy of Sciences but also upon its many predecessors. The first society that brought together scientists in the Czech lands was *Societas incognitorum,* active between 1746 and 1751. The oldest, truly long-existing (1773-1952) learned society was the Bohemian Learned Society (called Royal since 1784), which encompassed the natural sciences and humanities. Its founders included the eminent Czech philologist Josef Dobrovský (1753-1829), the historian Gelasius Dobner (1719-1790) and the mathematician and founder of the Czech University Observatory Joseph Stepling (1716-1778). In later years, the Society was led by the outstanding Czech historian Frantisek Palacký. As early as 1861, the famous biologist Jan Evangelista Purkyne (1787-1869) proposed in his work "Akademia" that a self-governing, non-university research institution be formed which would incorporate scientific institutes representing the main branches of science of that time. This vision of an institution devoted to interdisciplinary research was very close to the concept and structure of today's Academy of Sciences.

At the end of the 19th century, the country's scientific institutions were divided according to language: the founding of the Czech Academy of Sciences and Arts (1890-1952) was followed by the establishment of the Society for the Support of German Sciences and Arts and Literature in Bohemia (1891-1945). The Czech Academy of Sciences and Arts was founded thanks to extraordinary financial support from the Czech architect and builder Josef Hlávka, who also became its first president. The main purpose of the Academy, subdivided into four categories (humanities, natural sciences, philology, and arts & literature), was to support science and literature in the Czech language and promote Czech arts. Its most important activity was publishing. The Academy also granted scholarships and other financial support, such as an award for the construction of Jaroslav Heyrovský's polarograph, financing for Bedrich Hrozný's research trip to Asia Minor, and projects by many Czech artists. The Academy remained a learned society (its foreign members included such prominent figures as D.I. Mendeleyev and Marie Curie-Sklodowska), but minor research facilities were also set up within its framework.

After the creation of independent Czechoslovakia in 1918, other scientific institutions were established, such as the Masaryk Academy of Labour, and independent state-run institutes including the Slavonic, Oriental, and Archaeological Institutes. Lively international contacts cultivated by the domestic institutions made it possible for the Academy to join the International Union of Academies and the International Research Council.

After the communist-dominated totalitarian regime came to power in 1948, all the main existing scientific institutions and learned societies were gradually abolished and replaced by the Czechoslovak Academy of Sciences (1953-1992), which was both a learned society and a network of research institutes. Although science, in general, was exposed to crude ideological pressures until the downfall of the totalitarian regime in 1989, in many cases the Academy proved its creative potential and - to varying degrees in different periods - found its own path to the international scientific community. This was illustrated, for instance, by the awarding of the 1959 Nobel Prize in chemistry to Jaroslav Heyrovský, and the international recognition accorded Otto Wichterle for his invention of contact lenses. Of the other outstanding representatives of Czech science who worked in the Academy of Sciences in the past, at least the mathematician Eduard Cech, the theoretical physicist Václav Votruba, the geophysicist Vít Kárník, the physiologist Vilém Laufberger, and the philosopher and co-author of Charter 77, Jan Patocka, should be mentioned.

3. Science Policy

The establishment of an overall science policy is fundamental to the Academy's ability to fulfill its mission and make the best use of its material, human and financial resources. Science policy is formulated on three levels: first, within the individual institutes with the cooperation of the directors, scientific councils and scientists themselves; second, within nine clusters of institutes grouped together in similar scientific fields (the so-called "Sections"), where the proposed policies are synthesized and refined; and finally, within the Council of Sciences of the Academy, which submits the final version to the Academy Council and Academy Assembly for their approval. Science policy takes into account the current world-wide trends in various scientific (R&D) fields, the needs of the Czech society, conclusions of periodical evaluations of institutes carried out by the international Academic Appraisal Commission, and recommendations from universities, various institutions concerned with research and development, and governmental bodies.

Applying the principles of academic freedom and responsibility to the liberal environment of the institutes and Academy as a whole, the policy is not mandatory or directive but rather provides a framework for internal operations from the institutes to the top leadership. The implementation of the policy is based mainly on targeted and non-targeted financing, personnel policy, structural changes within the institutes, and differential support of individual research fields. Scientific competition on an international level is strongly supported. Science policy is not rigid, its formulation and application have a cyclic character and are re-evaluated and improved every few years in response to the world-wide situation and trends in science.

4. The Academy and Education

The Academy of Sciences has the status of a graduate school, where approximately 700 Ph.D. students work on their theses. In addition, many M.Sc. students participate in scientific programmes of the Academy's institutes.

These activities are mostly organized in collaboration with colleagues from universities. For example, the programme of doctoral studies in biomedicine is located in the Centre of Graduate Studies of Charles University and the Academy of Sciences; the Institute for Theoretical Study and the Institute of the Principles of Education are both joint establishments of the Academy and Charles University; the Academy's bio-ecological institutes and South Bohemian University share not only the campus but also the laboratories and other research and educational facilities.

Academy scientists give lectures at several Czech universities (30,000 hours in 1995) and teach in practical courses.

All of these activities are considered extremely important to the Academy: they help to recruit young scientists and graduates, and in return, they enrich university curricula in many fields.

Collaboration of Academy institutes and universities in research programmes is mostly mediated by collaborative grants of individual scientists and teams.

5. International Collaboration

International collaboration is an important and integral part of research and development. If appropriately conducted with the right partners, it provides a very effective way of solving research problems and promoting the goals of both the Academy and state science policy.

The research projects and programmes are formulated in agreements of three types:

- agreements between institutes of the Academy and research institutes or university faculties abroad
- agreements between the Academy of Sciences of the Czech Republic itself and its foreign partners, and
- agreements on science and technology or cultural matters signed on a governmental level.

In 1996, there were 230 agreements signed by institutes of the Academy, and 56 agreements guaranteed by the Academy of Sciences itself with its partners in 33 countries. They are devoted to very progressive research projects and development of technologies in the spectrum of research fields covered by the Academy and its institutes. The projects supported within the agreements are approved via competition.

Another important activity concerns international, governmental and non-governmental organizations. The Academy of Sciences has been charged with guaranteeing the collaboration of the Czech Republic with the following institutions: CERN - The European Laboratory for Particle Physics in Geneva; JINR - The Joint Institute for Nuclear Research in Dubna; the UNESCO programme MAB - "Man and Biosphere"; IIASA - The International Institute for Applied Systems Analysis in Austria; ELETTRA - a synchrotron radiation source in Trieste; the US National Science Foundation; the International Centre for Theoretical Physics of UNESCO in Trieste; the Czech Institute for History in Rome, and others.

The Academy of Sciences represents the Czech Republic in the International Council of Scientific Unions and is thus responsible for 35 national scientific committees in the Czech Republic under the umbrella of the ICSU. The Academy is also a member of Union académique internationale.

The Academy, via its institutes and research teams, participates in various EU research programmes (TEMPUS, PECO, COST, COPERNICUS, Fourth Framework Programme, INCO etc.), and closely collaborates with the NATO scientific committee and ALLEA (All European Academies). Moreover, the Academy of Sciences subsidizes international activities of 53 Czech scientific societies and organizes about 60 international scientific conferences or symposia per year.

STRUCTURE OF THE ACADEMY

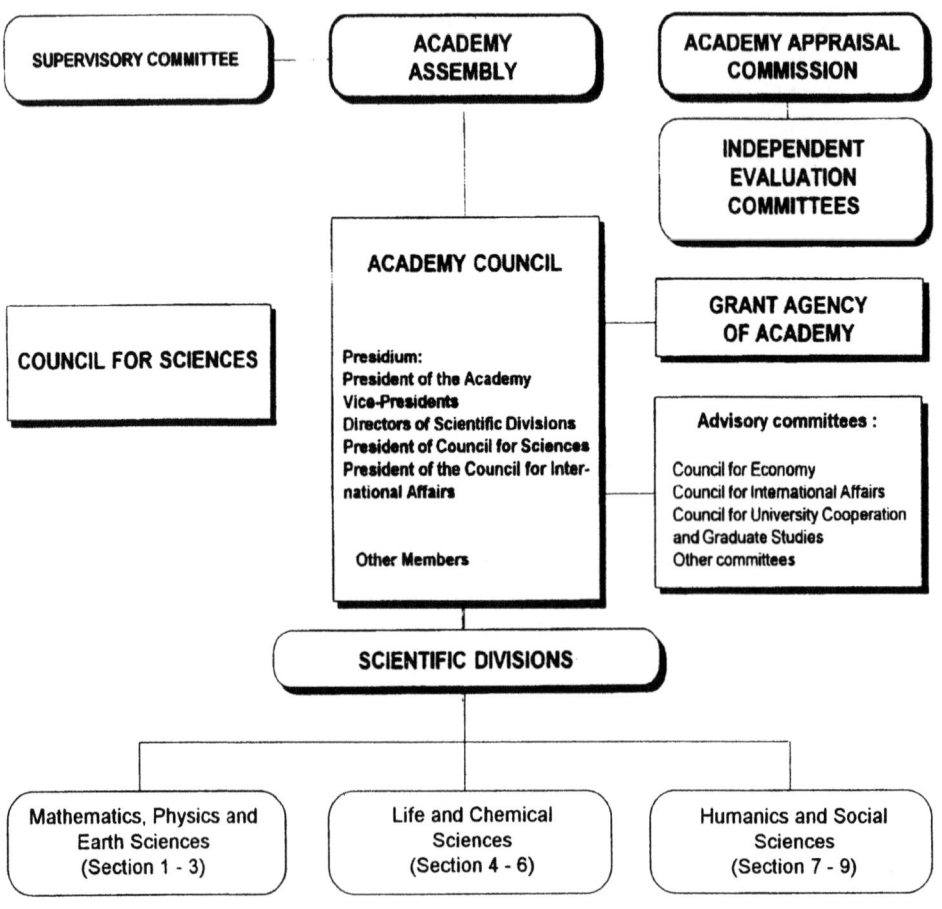

Nine Sections of the Three Scientific Divisions

1. Section of Mathematics, Physics and Computer Science

Astronomical Institute
Institute of Computer Science
Institute of Information Theory and Automation
Institute of Physics
Mathematical Institute
Nuclear Physics Institute

2. Section of Applied Physics

Institute of Electrical Engineering
Institute of Hydrodynamics
Institute of Physics of Materials
Institute of Plasma Physics
Institute of Radio Engineering and Electronics
Institute of Scientific Instruments
Institute of Theoretical and Applied Mechanics
Institute of Thermomechanics

3. Section of Earth Sciences

Geological Institute
Geophysical Institute
Institute of Atmospheric Physics
Institute of Geonics
Institute of Rock Structure and Mechanics

4. Section of Chemical Sciences

Institute of Analytical Chemistry
Institute of Chemical Processes
Institute of Inorganic Chemistry
Institute of Macromolecular Chemistry
Institute of Organic Chemistry and Biochemistry
J. Heyrovský Institute of Physical Chemistry

5. Section of Biological and Medical Sciences

Institute of Animal Physiology and Genetics
Institute of Biophysics
Institute of Entomology
Institute of Experimental Botany
Institute of Experimental Medicine
Institute of Microbiology
Institute of Molecular Genetics
Institute of Pharmacology
Institute of Physiology
Institute of Plant Molecular Biology

6. Section of Bio-Ecological Sciences

Hydrobiological Institute
Institute of Botany
Institute of Landscape Ecology
Institute of Parasitology
Institute of Soil Biology

7. Section of Social and Economic Sciences

Economics Institute
Institute of Psychology
Institute of Sociology
Institute of State and Law
Library
Masaryk's Institute

8. Section of Historical Sciences

Archeological Institute, Bmo
Archeological Institute, Prague
Archives
Institute of Contemporary History
Institute of History
Institute of the History of Art
Institute of Musicology

9. Section of Humanities and Philology

The Czech Language Institute
Institute for Classical Studies
Institute of Czech Literature
Institute of Ethnology
Institute of Philosophy
Oriental Institute

A STRATEGY FOR KNOWLEDGE DIFFUSION

E. MALITIKOV
International Association ZNANIE
4 Serova Proezd, Moskow, 101813 Russian Federation

Abstract

The intellectual community is joining efforts for no less than the sake of humanity. Changes are taking place in all spheres: political, economic, scientific, and others. Scientists of the former Soviet Union had the possibility of sharing their knowledge not only via different meetings and publications. There was a way of popularizing scientific achievements in the framework of "Znanie" (in English *knowledge*) Association. The aim of it was dissemination of knowledge among students, industrialists and other non-scientists. It was established in 1947 on the initiative of prominent workers in science and technology.

Recently it has been transformed into the international association "Znanie" headquartered in Moscow. It plays an essential role in the integration processes between the countries of the CIS, whose non-governmental scientific and educational organizations and those of the Baltic countries, still constitute the core of the association. The motto of this association is "Knowledge is the Bedrock of Accord, Security, Progress and Prosperity", its main aim is to promote science and education [1].

"Znanie" participates in the joint national and international projects under the aegis of UN, UNESCO and other international organizations. One of them is "I am the Man of the Earth". It relates to the sphere of social sciences and is aimed at the shaping of a new personality - a man of the future, and at the moral improvement of society.

1. Introduction

There is not much time left till the beginning of the XXI century. It is hardly possible to foretell what it has in store for mankind. But it is quite possible to draw some conclusions and learn some lessons from our activities in the current century. Although it is and has been the age of contradictions and

201

Ch. Proukakis and N. Katsaros (eds.),
The New Role of the Academies of Sciences in the Balkan Countries, 201–208.
© *1997 Kluwer Academic Publishers. Printed in the Netherlands.*

202

conflicts, the age of two world wars and of numerous local ones, the age of social and political cataclysms - still the XX century will remain in the book of history as the chapter proving the inexhaustible potential and might of science.

2. International Association "Znanie" - 50 Years of Experience

It is science, scientific and technical progress that became the symbols of our epoch, determined the current world situation, shaped the contemporary state of mind. It is not an exaggeration to say that crucial in these tremendous changes was the role of scientists, of their discoveries, and that of scientific organizations and communities whose aim was to popularize and disseminate knowledge. These organizations are traditionally referred to as "enlightening bodies".

One of the latter type is the International Association "Znanie". It is the successor of the former all-Union organization under the same name and it was set up in 1991 (see also [1]).

"Znanie" (this Russian word in English means *knowledge*) is a voluntary, non-governmental, non-political, non-commercial body uniting both national and international enlightening and educational public organizations. In 1991 all the educational societies from the former USSR (except Lithuania) became branches of "Znanie" Association, still constituting the core of it. The membership of this association has increased greatly in recent years: similar societies from Italy, Mongolia, the USA, Finland, the Czech Republic, Japan, China, Romania have joined it since 1991.

The 1993 Session of the Economic and Social Council of UNO admitted "Znanie" Association as an associate member of it. Today all member-states founders of the Association are officially recognized by the Economic and Social UN Council.

The "Znanie" Association stemmed from the All-Union Society. The latter was formed in 1947. So, next year will be a special year for "Znanie" - the society will mark its 50th anniversary. It will be a significant event commemorating the parent organization of today's "Znanie". Being a good child, the organization will pay due respect and tribute to its parent.

The association has accumulated a vast experience during these 50 years, put the cornerstone for many present-day traditions in this field. The experience. at the same time, is very instructive. So, what are the lessons that it was possible to learn during all these years ?

Lesson 1. It is necessary to constantly pay attention and support to enlightenment movements of every kind, and this is the task of the society and its governmental bodies.

Owing to that state care and support, the organization has matured rather quickly. It soon became one of the most authoritative and respected public organizations in the former Soviet Union. It was financed not only from the state budget, it had its own sources of finance, too. First of all, by giving lectures and publishing respective educational and popular scientific materials. But, certainly, state support was primary. Today, however, governmental assistance is minimal, which significantly lessens the effect and scope of its activities.

Lesson 2. It spurs the effect if the best members of the scientific community take part in the enlightenment and education of the population. It is in direct connection with the fact that for a lot of years "Znanie" was headed by world-known scientists. Among them were the Nobel laureates N.N.Semyonov and N.G.Basov, many holders of State Prize for Science and Technology (one of the highest distinctions in the former USSR). The USSR Academy of Sciences, university professors were active participants in the activities under discussion. It is sufficient to say that the first President of the All-Union Association "Znanie" was S.I.Vavilov who was also the President of the Soviet Academy of that time. Being a member of the association was socially prestigious; it was a mark, a kind of a label, of becoming a personality in scientific or cultural society of the country.

Unfortunately, now few people consider this work prestigious, especially young ones.

Lesson 3. Distribution, dissemination of scientific and cultural knowledge is not a work for a layman. This activity needs professionals, skillful workers, and a certain structural system. Most obviously, the core of it should be an all-national body, with branches in various cities. It should also include associations directly involved in enlightenment: councils of lecturers, publishing houses, its own mass media. This is what the All-Union "Znanie" once had and fruitfully used. Some of its periodicals are still being published. Among them such well known ones as: magazines (in Russian) "World Science", "Knowledge is Power", all-Russian newspapers "Business World", "Arguments and Facts", "Science and Business". Not long ago an annual collection of papers of leading scientists was published under the title "Science and Mankind".

Now efforts are being made to maintain and strengthen the already existing system. It was the strive for economic and spiritual revival of Russia and other Independent States that led to the joint project with the motto "Knowledge is the Bedrock of Accord, Security, Progress and Prosperity". It is the result of the cooperation of the Association "Znanie" and the Passport International Publishers in order to promote breakthrough innovation technologies and ensure a dialogue between the state and society on the topical problems of progress and security. "Znanie" maintains constant close contacts with UN, UNESCO, UNIDO and other international organizations.

204

Especially fruitful is the cooperation with the International Council and European Association for Adult Education. Quite recently, in August 1996, an International Educational Centre under the Association "Znanie" was established. Only during the last two years, the representatives of "Znanie" have participated in various forums on adult education, such as the General Assembly of the European Association for Adult Education (Copenhagen), the Annual Meeting of East-European member-states (Florence), International Workshop of editors of publications for adult education (Helsinki), regional workshops on enlightenment and distribution of knowledge in Bulgaria (Sofia) and Germany (Saltzburg). At present, the work is in progress for the World Congress on Adult Education to be held in 1997 in Hamburg.

An important event in the life of "Znanie" was the Forth Congress that took place in October 1996 in the Republic of Moldova (Kishinev). During that Congress, seven scientific and enlightening public societies from Russia and Romania joined the International Association "Znanie". Also a new conception for the Association activities was adopted. Under that conception, there are to be three main directions in further activities of "Znanie": continuous education, enlightenment, culture. Various projects are envisaged in every direction.

For example, in 1997 (the year of the 50th Anniversary of "Znanie") several international conferences are to be held: "Community of Independent States - a New Geopolitical Reality", "In Search for a New Harmony for Man", "Megapolises and their Role in the Evolution of Modern Civilization".

The International Association "Znanie" is open for new mutually beneficial contacts with those who share the basic principles of the Organization.

3. Some Present Problems of Science

It may seem a paradox that at the end of the century that put science and scientists in the centre of world developments, they are facing problems unheard of some 10 or 15 years ago. Below are several reasons, not all of them, that may account for the situation.

First of all, the effects and the role of science have become less attractive for the larger public. One of the explanations may be the following: Throughout the XX century the developed countries treated science as a fetish. But practice showed the inability of science to be a panacea for all social "diseases". It gave rise to various superstitions, mysticism, occultism, so-called para-scientific conceptions. In its turn, it made state governors become more cautious with the recommendations of scientists. It led to the uneasiness of many scientists who came to believe that "the Golden Age" of science was approaching its end, science turning into a tool for solving certain production tasks but not general social or political problems.

Thus, everywhere science and technology are losing their primordial function of a universal means for bringing the world to perfection, and are ceasing to be the most precious components of the world culture.

Secondly, there is the universal problem of environmental degradation, for which many people think science is responsible,. Not only the leading figures of "The Green Party", but also a number of well-known scientists, political and state officials are concerned with the drastic sharpening of ecological problems. They also express their doubt as to the ability of modern science to find a way out from a situation which science has created. These ideas are not prevailing in the society yet. But they have a tendency to be more widely spread. This dangerous trend should not be overlooked, because being accumulated these ideas may turn into obscurantism.

Thirdly, there is a problem facing mostly new independent states in Eastern and Western Europe, former Soviet republics: a lack of finances for promoting and developing science, both fundamental and applied. The current "poverty" of the scientific community in these countries is an extremely dangerous phenomenon because if it is not stopped, it may destroy scientific potential. Also, because a job in science is becoming less and less prestigious, less authoritative, this may lead to a spiritual decay of the society, nation, and country.

Two possible conclusions may be drawn here: 1) science now is at a crossroad, which requires changes in the attitude of society to science; 2) these changes in the attitude should not be to the extreme - either "scientism" or no science at all.

If we do not discard the idea of progress, if we consider that despite all difficulties mankind is going forward, not backward, then it is necessary to admit that the role of science and technology in this evolution is decisive. Any other inexhaustible resource than that of the human mind is hardly possible.

Science, in fact, does give rise to problems, sometimes very acute. Certainly, there is no direct link between science and morals, or between science and humanism. But these concepts are not antonyms. It is both possible and necessary to look for the ways to channel the best of scientific achievements into benefits for mankind. But it seems absurd to believe that any global problem could be solved without science, without scientists.

Therefore, one of the major objectives and obligations of the scientific community today is to pool efforts to re-establish and strengthen the leading role of science in society. To this end, two directions of activities appear to be suitable:

- vigorous involvement of science, alongside its increasing effectiveness, in the solution of global problems facing mankind on the threshold of the XXI century;
- realization of the integrating and educational aspects of science by disseminating scientific knowledge.

Both of these problems are concerned first of all with ecology. Ecology today is not only a system of respective knowledge, but also a system of certain values on which the survival of humankind may depend. So, ecological conceptions should be transformed into a kind of postulates, into a kind of specific faith (see more about ecological education in [2]).

Another aspect of today's life which demands heavy involvement of science is sustainable development [3,4]. The International Association "Znanie" is carrying on its activities in this field according to the basic principles offered by the UN Committee for Sustainable Development: development of education, informing population. training personnel. The Association "Znanie" is ready to take part in every project in any of these directions mentioned.

Three aspects are worth pointing out here. First, in Russia, as in any other economy of transition, the concept of sustainable development has not yet turned into a basic one for home policy. It is in the shadow of another, more acute current task - to find a way out of the economic and social crises, to stop decreasing the standard of living. Lack of investments sometimes causes taking measures that are against sustainable development. This divergence of practice from theory is an obstacle on the way to pertaining and implementing the concept of sustainable development.

Secondly, in Russia and in other economies in transition it is the state that plays the principal role in the solution of the problems mentioned. It is the state that has to take on the major share of the burden. Without state support, the efforts of the scientific community and the public are not enough here.

Unfortunately, due to the current aggravated economic recession, the government of Russia as well as the governments of other former Soviet republics have but restricted the possibilities to support activities and ideas aimed at the realization of the conceptions of sustainable development. In this direction, Russia lags behind a lot of states. A step in overcoming this difficulty may be the involvement of the International Association "Znanie" and of other similar organizations of scholars and scientists in the joint activities in this area.

Thirdly, so far, the public at large is not ready to take and promote the concepts of ecological education or of sustainable development. It is because of insufficient information that ordinary people, political figures and businessmen have on these matters. Hence, one of the major objectives of the Association "Znanie" at present is to bridge this gap. To this end, "Znanie" has elaborated its programme of introducing ecological education into the curricula at various educational levels.

4. Current Agenda of the Association "Znanie"

Along with the two very important issues mentioned above (ecological education and sustainable development), the dissemination, diffusion of scientific knowledge remains the core activity of the educational association "Znanie". Enlightening, educational efforts of scientists are becoming even more significant now, turning into one of the mightiest promoters of the integration of the mankind, especially so because the world appears to be more open and intra-related.

There is another reason why enlightenment in its widest sense comes to the foreground again. This is the response to the emerging of the ideal and political movements whose principles contradict the ideas of intellectual, spiritual and personal freedom. Those ideas are sowing seeds of ethnic, religious and political discord. They are: the extremist wing of the Islamic fundamentalism, radical nationalism, fascism with all its faces, and the like. All of them are dangerous not only for the world as such, but also for world science. To cease those movements, it is necessary to pool together not only by political but also by moral means, the dissemination of scientific knowledge.

A concrete contribution of the Association "Znanie" in this field is the project "I am the Man of the Earth". This project has been worked out in Russia and it is currently being implemented by the Association "Znanie" (see also [1,2]). It is a joint international project. Its goal is comprehensive evolution of a person. It refers to the so-called "social technologies" and consists of a number of sub-projects.

To make this project a really international one, the Association "Znanie" has currently begun the formation of an International Coordination Committee. It is now seeking the support of authorities in science and of prominent public and political figures from all around the world.

An important aspect of it is to design new ways and methods of enlightening educational activity. Certainly, such conventional means as giving lectures, organizing courses, publishing magazines and booklets, making audio and video recordings - are all valid in the sphere under discussion. But their great disadvantage is that usually they may be used by a limited number of readers, listeners or viewers. Modern global computer networks have nothing of this kind. Nothing of this kind have modern global computer networks. However, they are now very seldom used for education. But their role in the contemporary life is so great, that *one of the next NATO AR Workshops may be devoted to the role of modern information and telecommunication means in the process of scientific knowledge dissemination.*

5. Conclusion

Science, as well as our world as a whole, is passing through a rather unstable, turbulent period of its existence. But despite all troubles, science still remains the major spiritual resource for the progress of mankind. That is why the activity of every scientific community, including that of the Association "Znanie", is of great importance and value nowadays. Scientific organizations have a promising future, if every party concerned makes its contribution.

References

1. The intellectual community (editorial) (1996), in *Passport to the New World*, July/August, p.12.
2. Malitikov. E.M. (1996) Environmental education and environmental issues, in S.Radautsan, G.Parissakis (eds.) *Scientific and Technological Achievements Related to the Development of European Cities*, Kluwer Academic Publishers, Dordrecht, pp. 273-278.
3. Andriesh, A. (1996) Scientific achievements related to the sustainable development of cities, in S.Radautsan, G.Parissakis (eds.) *Scientific and Technological Achievements Related to the Development of European Cities*, Kluwer Academic Publishers, Dordrecht, pp. 13-22.
4. Radautsan, S. (1996) Science policy and problems of investment for the development of cities, in S.Radautsan, G.Parissakis (eds.) *Scientific and Technological Achievements Related to the Development of European Cities*, Kluwer Academic Publishers, Dordrecht, pp. 47-62.

ACADEMIES FROM PLATO TO ZAMOLXIS

A. VARTIC
Cultural Foundation Basarabia
18, Sfatul Tarii str., Kishinev MD 2012, Republic of Moldova

Abstract

The communication contains the author's ideas and hypotheses about an ancient school of knowledge not widely known yet - that of Zamolxis. He, as his doctor said, was either a King or a God of the Dacians. Those ancient people once lived on the territory of contemporary Romania and the Republic of Moldova. Parallels with Greek history are made. Unique archaeological findings of the author and his team on the territory of the Dacians and their capital Sarmizegetusa, are described. Conclusions are drawn relevant to the contemporary search for "harmony" between science and society, the role of academies in this search being decisive.

1. Introduction

The name of Plato, a philosopher from ancient Greece, and his Academy are familiar to every scientist. See details in [1]. But, most probably, the name of another great personality - Zamolxis - is not as widely known. Around 432 B.C., in Tracia, there was another philosophical school, or better said, another Academy, that of Zamolxis, who was either a God or a King of the Daco-Getae [2]. (As this communication is the continuation of the research into the subject of ancient people on the territories of contemporary Romania and the Republic of Moldova, for more information about these people, the reader is referred to the previous publication of the author in the NATO ARW Series [3]). This time the author addresses several new aspects of Dacian life in those times, with certain references to Greek history.

My personal acquaintance with Greece and Greek myths began with Helen-the-beautiful. And I understood, even when a boy, that love or hate are the motivating forces of great many events in history. But now I am not going to indulge you into a long story of Greek history. Being a guest in Greece, it would be slightly impolite. I would like to draw your attention to some facts

Ch. Proukakis and N. Katsaros (eds.),
The New Role of the Academies of Sciences in the Balkan Countries, 209–217.
© *1997 Kluwer Academic Publishers. Printed in the Netherlands.*

and episodes from the ancient history of my people, whose civilization is not as widely known as the Greek one, but maybe will begin to be better known from now on because the contribution of the Dacians in many scientific and technological discoveries was undeservedly obscured.

2. About Zamolxis and not only

In discussing the role of academies, one cannot avoid mentioning very often the name of the first "academician" Plato because not only his ideas but also his way of popularizing them is so much different from many contemporary purely scientific sophistications. Plato in "Harmides" graciously tells us about Socrates and his reflections on and arguments for Beauty, Beauty as a whole, and Beautiful utterances. (The given communication is also under the influence of aesthetic and artistic works by Borges, maybe because he was also "affected" by Plato). Plato tells us about Harmides, a nice fellow, who had a terrible headache, and one mysterious doctor, who was also the doctor of Zamolxis and was said to know how to make people immortal, noticed that Greek doctors could not cure many diseases [2]. That doctor used perfect scientific language in the discussion with the famous philosopher. So, it is possible to admit that there was a philosophical school, or maybe even Academy, of Zamolxis in 432 B.C. [4,5]. That doctor of Zamolxis knew that the whole body should be treated but not just separate parts of it, soul being the nucleus, the core of the "whole". (Here, perhaps, the name of Freud may come to mind because that doctor advised to politely ask people about their dreams). Socrates, the other partner in the conversation, noticed that to treat and cure the soul sometimes very simple but very beautiful utterances are enough.

Another evidence for the existence of Zamolxis's school can be found in Jordanes's works [6]. Those who would like to know more about Zamolxis and his school that was once located in Romania, in the Carpathians, are invited to read our earlier books on the subject. This is the direct way. Because following "Ariadna's thread" from Helen, the wife of Menelai, we came to Socrates and Plato, and then to Zamolxis, the latter being the King of the Dacians and their God, and at the same time a slave of Pythagoras. Falling in love with Helen-the-beautiful, I came to love ancient Greek history, as the history of mankind, then the history of my people. Indeed, beyond the grasp of the mind are all the ways to cognition. That is why now it is so important to study the two philosophical schools, or, better say, academies of knowledge, in parallel. One is based on the ideas of Plato (and also of Pythagoras, and of Thales) and the other came to be known from the archaeological findings of the Daco-Getae who lived in the times of Zamolxis.

3. The Dacians of Zamolxis

A lot is known and a lot is said about ancient civilizations: about the pyramids of Egypt, about Stonehenge, etc. Everybody knows about them from school age. Unfortunately, the Dacian civilization of the times of Zamolxis is not well known yet, even in Romania, on the territory of which a great number of the vestiges of that civilization were, and maybe will be, found. So-called citadels have been recently discovered that surround almost inaccessible mountain summits. They are arranged in either recognizable or strange geometrical figures, built of huge monoliths of 3-5 tones, brought from far-away places. There are also some strange geometrical figures near the citadels, only 5-10 cm above the earth, that remained practically intact. There are also a lot of miracles connected with these citadels and sanctuaries. Moss grows there but a little. In general, biological-organic life is very weak there and very scarce. To give only one example of the capabilities of the Dacians, it is worth mentioning here that they constructed thousands, tens of thousands of perfect horizontal stone terraces high in the mountains. What is even more important is that metal slag is found practically everywhere on these terraces. It seems that the people mentioned by Socrates knew metallurgy. And maybe, why not, they were the best metallurgists of their time. Romanian archaeologists have found metallic magnifiers of 40 kg on the outskirts of Sarmizegetusa, the Dacian capital. (As it is known, ancient craftsmen could melt iron into magnifiers of 20-25 kg). Moreover, analyzing some so-called Dacian nails, we came to the conclusion that these people could get λ-iron of 99.97% purity. As to building materials, they used specific porous concrete doped with unusual metal alloys. The basic element of that concrete was not calcium. They used burnt sand treated with metallic compounds containing Ti, Ni, Va, Ag. These concrete terraces stood the test of time, of earthquakes in the area of high humidity, of high temperature. One more characteristic of the concrete - the surface exposed to humid air was covered with a thin film that could resist even acid.

The Dacians lived in almost perfect harmony with nature and space, according to their own solar and lunar calendar, which we tried to decipher. Here are some of the hypotheses offered. One of our main "tools" was the Hellenic "language", the "language" of symbols and parallels, the language of messages from antiquity. We, Romanians, also have our own Iliada that is, in fact, the book "Miorita". It is a ballad telling us about a sacrifice at sunset of one of the bravest and one of the richest men of that society. Herodotus also mentions sacrifices. The father of history says that once in five years the Getae chose the best of the fellows to sacrifice him as a messenger from Zamolxis to other Gods. These sacrifices are supposed to have taken place on a mysterious "solar andezite" which is still in a very good condition and can be seen in Sarmizegetusa, the ancient capital of the Dacians.

"Miorita" tells us that the ceremony of sacrifice took place at sunset. Hence, the starting idea of ours was to watch what is happening in that place on June 21st, the day of Summer Solstice. We went there in June 1996. The investigations of my team and myself confirmed the fact from "Miorita" that the witnesses of the sacrifice were just the Sun and the Moon. This is the illustration of what we observed:

The Moon	The Sun
9:05 p.m.	9:05 p.m.
June 21, 1996	June 21, 1996
$* a$	$* b$
$* c$	$* d$

Here points a, b, c, d indicate the location of the stones in one of the Dacian sanctuaries on an elliptic top terrace in the "citadel" near Costesti. The Sun reached the line d-b and the Moon d-a. This cosmic event, in good agreement with the Metonic cycles (Meton was another Hellenic genius, a 5th century B.C. astronomer), takes place only once in 19 years. The next one will be in 2015. The results of the observations (watched and registered by various photographers and cameramen, not only ours) are the following: the same event was first observed in our era, i.e. A.D., most probably on June 21, 1 A.D. Using these conclusions, it is possible now, using scientific graphs and diagrams to state when the Moon was "drawn" to the orbit near the Earth. A hypothesis of ours is that it was a result of a tremendous melting of ice that covered a larger part of the Earth in those times. Back to our points. On the line d-c, an old Daco-Getaen citadel, Piatra Rosie (Red Stone) is situated, on the diagonal b-c, perpendicular to the point of the appearance of the Moon according to Meton, Blidaru and Banita are located (this hypothesis is confirmed by the modern topographic maps). Moreover, Metonic cycles demonstrate another thing. Probably we happened to find the key to the old message in "Helen" language. It suffices to give one example here: using these key-algorithms, it is possible to ascribe number X to the word "Pythagoras". The word "Burebista" (the name of the most famous Dacian King) gets number 10X. This method is based on Metonic cycles and can be easily checked and verified. The wonder of some figures discovered by Meton in 432 B.C. is directly or inversely proportional to the principal syllabus EN which, in our opinion, is the basic element of the "Helen" language, or better say the language of Helen-the-beautiful.

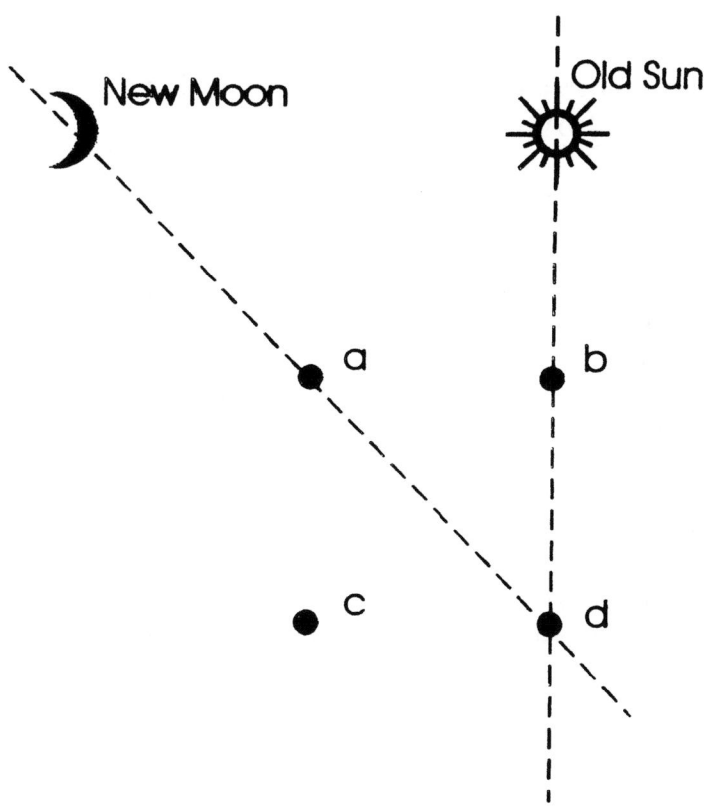

4. Laws of languages as laws of nature

The laws of language, as any other laws, including those of nature are to be found and studied. Now the time has come to understand what was not under the effect of radioactivity in "Feasts of Immortality" [5] organized by Zamolxis, a slave of Pythagoras, in underground chambers. They were probably held underground because on the surface of the territory once inhabited by the Dacians there were many radioactive stones. Those Dacians also left us some metal works of perfectly pure iron, superb groutings, palladium - a mixture of iron, titan and nickel with burnt wheat in Sarmizegetusa, exciting constructions and terraces in Fetele Albe (White Faces), Blidaru, Piatra Rosie (Red Stone), Virful lui Hulpe. In the Carpathians, on the Romanian side, there are some unusual mountains named Timpul (Time), Omul (Man), Steaua Mare (Big Star), Poiana Omului (the Glade of Man).

The Indian Panini once said that the effect of a word is not multiplied if we change the order of syllables. This might mean that the effect of EN is not stronger than that of NE. In earlier works of the author [7, 8, 9, 10], it was shown that the names of many places in Romania, where the Dacians once lived had the combination EN. Another large group contained names with ST (ShT in English), e.g., Milesti, Costesti, Bucuresti, Hincesti. What relation can there be between EN and ST? Panini helped us a lot in answering this question. First of all, it was he who inspired us to investigate AR, which is a constituent of the tribe's name Aryans, to which Panini belonged. The Aryans are known to have been a tribe of tillers. They had ploughs, horses, they sowed and harvested grain, threshed and milled it. They knew Vedas, those mysterious texts considered to be God's messages to the people on Earth. The key syllables of the words in Vedas are "om", "mana", "dova"; the God's names were Deva, Vishnu, Crishna; the divine drink was soma. To make a parallel with Romanian names, the name of the city Deva comes first, then the rivers Crish, Ariesh, Vishe, Somesh, Mana. The sacred syllable-word "om" is not necessarily translated in Romanian. It means the same - "man". The old capital of the Dacians was Sarmizegetusa, where Sar might have lived. The letter S symbolized a divine serpent. In old Egyptian drawings, God Ra usually had a nimbus in the form of a serpent. God Ar of the Aryan tribes might have lived in Sar, as the Semite God Ur lived in the town Ur.

Another name which comes to mind is Ghilgamesh. No one can say now if there is any link between him and Zamolxis. As Panini said, "om" is not multiplied if it is pronounced "mo". But this sacred combination "mo" is a constituent of the name Zamolxis. One of the rivers in Romania is Moldova or Mo-dova. That river gave the name to a large district in Romania and even the name of the country we are from - Republic of Moldova. Thus, "om", "mo" and "dova" had in the Sanskrit language the same meaning: it might have been the name of the same thing but seen from different points of view. Alas, it is not the time and point here to discuss any more fascinating and intriguing mysteries of old languages.

5. From Plato and Zamolxis to nowadays

It seems reasonable here to return to Plato whose Academy sent an important message across many years, which we cannot ignore. A man-of-science should not forget about politics if that *homo sapiens* would like to have a future on the globe.

Marcuse, Sartre, Foucault demonstrated it rather evidently. It should be forbidden to work on the creation of a new bomb without taking into consideration its possible social and political "after-effects". Thus, philosophical aspects become no less important than the investigation itself. Because evolution, as a concept of universal development, is an irreversible self-consumption, an amplification of entropy. Those who do not understand that "nice utterances" of the doctor of Zamolxis, the doctor mentioned by Plato in "Harmides", are all cognition, those who do not see that art, poetry, Vedas, Iliada, Ghilgamesh - are all cognition, they only magnify people's sufferings through their fundamental investigations. They, first of all, act against themselves. With these maxima of Solomon, I do not insist that scholars should first study Iliada, then become MPs in their countries, and only after that to get involved in studying the properties of uranium, for example. In antiquity, people understood better than many of us today that poetry is also cognition, maybe one of the major ways of cognition. That is, probably, why an ancient Greek sent us orally one of the most mysterious messages - Iliada. That gigantic book which, perhaps, even Shakespeare could not have written, tells us about a terrible world war caused by an abduction.

Today's tragedies, bloodshed, catastrophes caused by wars, provoked and waged by people in this dramatic XX century of ours - are all eloquent evidence of the great responsibility shared by the society and men-of-science. It became fashionable for the "white collars" from laboratories not to take seriously the ideas of Plato or other philosophers who, in the opinion of the former, are "great loosers" in science because they were involved in politics. The un-involvement of contemporary men-arts and science proves that, unfortunately, Solomon was right. Science strongly increases people's sufferings. So, the new role of the contemporary Academies in the Balkans, the region which provoked wars not only in antiquity but also two in our century started from the Balkans, is clear: a man-of-science has to be involved in politics. It is a duty of a man-of-science, as well as of a man-of-arts, to show political figures that we are all but tiny pieces (or, maybe, holograms?) of the Universe. And, as Plato once said, the Universe, one day, may stop liking our irresponsible tricks and exercises of different kinds.

This is the important message that the Daco-Getae of Zamolxis's times tried to transmit to us. According to Jordanes [6], the Getae knew that the Sun is bigger than the Earth. Other people of antiquity could not have known that the Sun is bigger than the Earth (as was much later proved by Copernicus). But, surprisingly, the Dacians knew *how many times* the Sun is bigger! They still lived following the laws of nature, which was, most probably, very important. In this, the Dacians were superior over other ancient people because they knew logic, and even science had to be in accord with nature. The Roman Emperor Traian covered this belief of the Daco-Getae with dust and sand, waging the war against the Dacians. The Academy of Zamolxis was a school in the world history for about 2000 years. That one and the Academy of Plato, though mocked at, especially in our century, stood the test of time. In fact, a battle with Plato's ideas is the battle with harmony between man and Universe, and the real wars in which all of us suffer, are a vivid result of that battle.

6. Conclusions

Here, in Athens, people know very well what tragedy the virus "Medeia" can cause. Contemporary scientists practically everyday go into so-called "expeditions in search for the golden fleece" into the yet unknown corners of cognition. With the advent of computers and their virtual realities, these "expeditions" might be converted into an incurable virus. A modern scientist does not filter these expeditions for the unknown, neither through "the siege" of Homer nor through the one of Orpheus. As we all know, there are no well-based programmes to combat the virus "Medeia", which as a parasite chooses the golden fleece as its host. Jason, another ancient hero fathered the syndrome of greed. But *greed for knowledge* is a very good thing. Unfortunately, the concentration of population in megapolises, breaking the harmony of man and nature (in other, more poetic, words, the domination of UR over AR), magnify violence, sufferings, natural disasters, various catastrophes. That is why the basic idea of the ancient Greeks about universal harmony is to be vigorously advertized, if we all, as mankind, want to live in the future. The role of the Academies of Sciences in this battle for survival on Earth, which is still a bit of a paradise, is crucial.

References

1. Theocaris, P.S. (1995) The Academy of Athens and its role on the cultural and technological development in Greece, in G. Parissakis and N. Katsaros (eds.), *Science Policy and Research Management in the Balkan Countries,* Kluwer Academic Publishers, Dordrecht.
2. Plato (1976) *Works* 2, science and Encyclopaedia Publishers, Buchuresti (in Romanian).
3. Vartic, A. (1996) Urban problems across two millennia, in S. Radautsan and G. Parissakis (eds.) *Scientific and Technological Achievements Related to the Development of European Cities,* Kluwer Academic Publishers, Dordrecht, pp. 303-313.
4. Herodotus (1964) *Histories* IV, Editura Stiintifica, Bucharest (in Romanian).
5. Vartic, A. (1994-95) Feasts of Immortality, *Quo Vadis* 1-3, Basarabia Publishing House, Kishinev (in Romanian).
6. Jordanes (1960) On the Getae Origin and Deeds, Moscow State University Publishing House, Moscow (in Russian).
7. Vartic, A. (1995) From monographic to entropic sociology, *Saptamina,* 2 June (in Romanian).
8. Vartic, A. (1996) *Letters to Bill Gates,* Basarabia Publishing House, Kishinev (in Romanian).
9. Vartic, A. (1996) 432, *Quo Vadis,* Basarabia Publishing House, Kishinev (in Romanian).
10. Radautsan S., Vartic A. (1996) The Antic civilizations and "Third Wave", Workshop World Watch, Snagov, Romania, March 22-23.

THE ROLE OF THE NATIONAL ACADEMY OF SCIENCES OF ARMENIA IN CO-ORDINATION OF FUNDAMENTAL RESEARCH AND SCIENTIFIC NETWORKING

YU. SHOUKOURIAN
National Academy of Sciences of Armenia
Division of Physical-Mathematical and Technical Sciences
24,Bagramian ave., 375019 Yerevan, Armenia

1. Introduction

In the transition period from the planned economy to the free market economy in Armenia, the basic research development policy cannot be treated only in the context of market rules [1], while applied research sometimes can be adapted to the market economy. The preservation and stabilization of fundamental science is a policy issue of primary importance. There are some new problems related to the co-ordination of fundamental research and scientific networking in Armenia, and the National Academy of Sciences is developing approaches for their decision.

Difficulties of the transition period in Armenia brought to a drastic reduction of the science budget. At present, there is only project based funding of science which was introduced in 1991 and was oriented to scientific teams which had already achieved a professional level. Though this type of funding has some positive features, the lack of the institutional (basic) funding and priority research programs [2] has put under threat the vitality of scientific institutions and of the scientific environment.

The research institutions still have experimental equipment which frequently are unique and their restoration on the international level in the near future is hardly by means of state funding. This becomes a very important problem for those fundamental investigations which require an efficient experimental basis. There are also good possibilities for co-operation between scientists from different countries in the collective use of unique equipment. Another problem is to establish new branches of innovative research addressed to the needs of national economy and society, and to create a unified information space for science and education.

At the same time, Armenia faces also the typical for NIC ''brain-drain'' and ''ageing'' problems. It is extremely important to select gifted

219

Ch. Proukakis and N. Katsaros (eds.),
The New Role of the Academies of Sciences in the Balkan Countries, 219–223.
© 1997 *Kluwer Academic Publishers. Printed in the Netherlands.*

young researchers by organizing a special state supported program for the support of young scientists. There is some progress in the creation of telecommunication links to Internet, which gives a chance to fill the "information vacuum". Simultaneously, the process of developing an internal data communication network is advancing very slowly. There are many problems related to the translation and publication of academic journals.

2. Co-ordination of Fundamental Research

As the co-ordinator of fundamental science in Armenia, the National Academy of Sciences has initiated some new steps in the transition period.

The first activity of the Academy is to establish the priority areas for basic research, taking into consideration the existing science base, current needs and situation in Armenia. In the last years, nine problem-oriented scientific councils have been established. These councils are uniting representatives of academic research institutes, branch institutes and university science in the country. The main area for activity is the development of science policy, creating recommendations to the government and defining priority areas of science.

The second phase is to create a list of unique scientific equipment, available in the country, and present it to the international scientific community to generate joint research projects and co-operative programs. Such apparatus must be of a national importance and can be available for use by researchers from different countries. There are possibilities for co-operative arrangements on an international level and for the establishment of joint research centers. The Armenian Academy of Sciences has some experience in this area (astronomy, physics, biotechnology, geophysics and geology).

Thus, for more than 5 years, the observations at the 2m60 telescope in Byurakan astrophysical observatory were interrupted because of economic difficulties. But such observations are of great interest for researchers. In 1995, a joint French-Armenian astrophysical meeting took place in Byurakan to define the new fields of co-operation. It was suggested to put again in operation the 2m60 telescope, which is one of the largest in the world. For this, the French side will build a special multi-purpose instrument to be attached to the prime focus of the telescope. This instrument gives efficient observations in imagery, interferometry and field spectrography. The telescope will be able again to get high quality observational data in domains of strong international competition. Efforts of both communities were considerable. So, the large, high level telescope of the Byurakan Observatory, which has not operated for 5 years, recently received its first new observational materials.

The traditional key objective of any academy is to establish new branches of fundamental sciences through priority oriented programs and innovative research addressed to the needs of national economy [3]. In the transition period, it is very important to have institutional and project-based funding. Clearly, difficulties related to the transition period do not give possibilities for a wider support from the budget. But this form of funding is an essential source for the creation of new scientific branches. The problem of setting up joint ventures by different institutions of Armenia can also be solved in this way. In the last years, Armenia suffered from a catastrophic earthquake (1988), continued blockade, energy shortages etc. To respond to new needs, the Armenian Academy developed proposals for scientific investigation of the problems of emotional stress, involving a complex of medicabiological aspects: molecular mechanisms of neurochemical and cellular activity, placticity of the neural system, change of immunological and psychological activity, influence of stress on the mother and child development, creating of antistressary drugs, etc. This requires to attract the attention of fundamental researchers and can create the possibility for national collaboration. There is certain progress in the collaboration through the INTAS and NATO Science Program. There are very important problems for humanitarian sciences: armenistics, economic sciences.

Meanwhile, Armenia did not avoid the "brain-drain" and "ageing" problems and now it is extremely important to select gifted young scientists and involve them in basic research, organizing a state supported program for the protection of young scientists.

3. Scientific Networking

The creation of a contemporary information infrastructure for the Armenian scientific community was an urgent necessity during the last 4-5 years [3]. Presently, essential progress in linking with the outside world via international networks has been achieved. Important steps have been taken in the formation of a scientific telecommunications network, which will enable us to have access to Internet via the satellite telecommunications node of the Yerevan Physical Institute.

The progress in linking some basic nodes (scientific institutions) through wireless subnetwork with access to Internet enabled to involve more than 15 academical institutions in network. The problem of the "information vacuum" for the scientific community of Armenia has been solved due to the support of government, the Foundation "Armenia" and the Armenian Diaspora. The NATO Networking grant, which started in 1996, gave a possibility to increase the number of users. Some academic institutions are sufficiently far from Yerevan. For this reason and to integrate them by the existing nodes in the Armenian Scientific Data Exchange Network, a network for the information transfer for research institutions and research organizations will be created, opening access for the Byurakan - Ashtarak group of institutes to the Internet. This is the continuation of the NATO networking project for Armenia. The Byurakan Astrophysical Observatory (BAO), which has access to Internet via radiomodems fragment, will become the basic node for this region. At present, it is necessary to define the strategy of developing the scientific information resources of Armenia. The guidelines for it are the creation of a unified information network of science and education in Armenia and active usage of the international networks. To solve the above mentioned problem, the following steps should be taken:

- To develop the national database on Science and Technology, using it as the background for the extended international co-operation. It is very important to develop existing WWW-servers of the Academy, Yerevan Physical Institute, State University Engineering and to create new servers and home pages.
- To create training and educational centers for scientists and students. It is necessary to diffuse the contemporary methods for efficient use of network services.
- To start the pilot project for the transition of voice and image technology for scientific investigations, developing Armenian language engineering and creating a media for multilingual word processing.

4. Conclusion

In the period of transition from the planned economy to the free market in Armenia, there are some new problems which are related to the co-ordination of fundamental research and scientific networking: generating joint research projects on the base of unique scientific equipment available in the country; establishing new branches of innovative research that is addressed to the needs of the national economy and society; creation of a unified information space for science and education.

References

1. Sarkissyan, Yuri L. Guidelines For Science Policy and Research Management in Armenia, in *Science Policy and Research Management in the Balkan Countries,* 165-171, 1995, Kluwer Publishers, Netherlands.
2. Plompen, A.P. Research Management in the Netherlands, in *Science Policy and Research Management in the Balkan Countries,* 141-155, 1995, Kluwer Publishers, Netherlands.
3. Shoukourian, Yu. Information Technology Transfer in Armenian Science, in *Science Policy and Research Management in the Balkan Countries,* 173-177, 1995, Kluwer Publishers, Netherlands.

SCIENCE AT THE EDGE OF THE TWENTY FIRST CENTURY:

Social Results and Problems

YURIJ SOLODUKHIN
Moscow International Association "ZNANIE"
40, Frunzenskaya Naberejnaya ap. 15, Moscow Centre, Russia

In my speech, I would like to touch upon some common aspects of science condition, scientific institutions and communities in the modern world.

The fact of undulating development of science is unrefutable. Looking at the three thousand year old history of science (as this very period is properly documented), we see the replacing of revolutionary rises and depressions, each rise in different fields of the natural history and the humanitarian problems could be asynchronic.

We live in the epoch of a creative and scientific and technical boom, which started at the edge of the 19th and 20th century.

I do not intend to give the full list but will stress only the peculiarities of the current period which made it possible to get the qualitatively new role of science in the modern society, to get new forms of organizing scientific research, to get the new type of relations between science and state and science and society.

This is, at first, the duration and the universality of the scientific and technological revolution: it has been lasting for almost a century and it covers all the main spheres of science and technology. As mentioned above, the undulation reveals in fact that for a century the leading role from the point of view of the revolutionary character of a discovery has been taken in turn by different sciences and their divisions (physics-theoretical, nuclear, solid, etc., mathematics, genetics, chemistry, cosmology, informatics). To contrast the previous centuries, our century didn't have depression periods, a scientifical or technical stagnation. The 20th century is the century of permanent scientific and technological revolution.

The second feature is the sharp reduction of time between a scientific discovery on the one hand and its technological realization and practical usage on the other hand. The very scientific achievements and ability to make them practical advanced technologies in a rapid and rational way have

225

Ch. Proukakis and N. Katsaros (eds.),
The New Role of the Academies of Sciences in the Balkan Countries, 225–230.
© *1997 Kluwer Academic Publishers. Printed in the Netherlands.*

become the main resource of movement forward for a country, nation and the factor of great influence on the domestic and foreign policies.

A kind of politicization of science took place. Not only in that positive meaning that governments, state, society recognized the social consequences of scientific and technical leaps and took into consideration the position, opinion and expertise of scientific community. A negative politicization of science also took place and revealed in usage the benefits of science as a tool for achievement in a unilateral national or bloc exceed in economical, political or military sphere and also in usage for manipulating behavior and thinking of people.

The third feature is the transformation of science into labor-intensive and capital -intensive sector of social life. This is an objective process that can't be avoided. But this made scientific development and scientific community dependable on both state and private sources of finance. There appeared a contradiction between the spirit of intellectual, creative freedom which is the essence in life of a science and a scientist on the one hand and an aspiration to influence the process, to keep it under control revealed by power and also by a science-financing business.

And, last, there is the fourth feature which I would like to dwell into detail as it characterizes, to a great extent, both the role of science in the modern world and the attitude of society to it. I mean that two kinds of global consequences of science and technological leap:

- on the one hand, this process demonstrated in the twentieth century colossal abilities for mankind by a scientific reflection and technologies, based on it. It helped, in the first place, to provide a really high living standard and comfort for a meaningful part of the population in many countries. A persuasion was formed in the society that science is able to overcome all problems, including those of a humanitarian and social character, by means of using methods of science mostly of natural sciences. This was reflected in the so called scientism, in the constructing of social engineering, in political technologies;

- on the other hand, in the very twentieth century the world realized that the development of science and technology brings not only colossal abilities of prosperity but also real threats to the survival of mankind. These threats are affected by the creation of weapons of mass distraction, at first nuclear weapons, and also with sharp deteriorations of the environmental problems. However, as the constructing, spreading and usage of weapons can be controlled and liquidated step by step, the ecological situation requires absolutely new approaches. In the first place it requires forming a qualitatively new thinking, a view upon the correlation of nature and mankind, spiritual and industrial activities.

It is inefficient to be sorry about the lost time and possibilities. One can't help remembering the outstanding scientist Vernadskij who wrote in 1926 in one of his articles: "Biosphere begins changing much stronger and deaper under the influence of scientific thought of mankind. A newly born geological factor - scientific thought - changes life phenomena, geological processes, the planet energy. It's obvious that this side of scientific trend of man is a natural phenomenon". Unfortunately, the concept of a steady development based in the essence on the quoted philosophical purpose became the property of world community only sixty years later.

What critical conclusions are drawn from the above?

1. It was noted long ago that the scientific leaps are directly attended by the concentration, in one or some generations, in one or some countries, of the people whose talent or genius provide breaks to the new horizons of knowledge. For instance, the revolutionary remaking in the theoretical physics in the end of the 19th and 20th century could hardly take place but for the living and working of M. Plank and A. Enstein, N. Bohr and V. Geizenberg, P. Dirac and I. Tamm, L. Landau and P. Kapitza, A. Alexandrov and S. Ambartsumjan at the same time.

The appearance of such concentrations, pleiades of the outstanding persons both in science and culture cannot be explained neither in genetic nor in social and political terms. Society cannot influence this process.

But geniuses do not appear in the blank space, they arise from their predecessors. In science, such predecessors are the so called scientific schools. They embody exactly the succession, accumulation and the followed development of scientific ideas and trends. These are long term orients and tools of advance, movement of science. With the end of their existence, rich layers of scientific ideas and knowledge fall into decay.

It cannot be tolerated. I think in its concluding documents, our congress should stress the necessity of consideration of preserving and developing the leading scientific schools as a prior tendency in the work of planning and financing scientific researches. We should persistently recommend this to governments, ministries of science and technology, higher education, academies of sciences, non-governmental organizations which influence on science in this or that way.

2. Science, especially fundamental science needs a permanent and sufficient financial support from state and other sources. If it is not done, if science is forced to come to self-finance there will be a warp in its development to applied research and today's leap of scientific thought will change into depression.

This threat is real for science and scientific institutions in Russia. The Science Academy, its departments, other scientific institutions of the country adjust to work in new conditions, learn earning, find financial sources alternative to the ones of the state. But these opportunities are limited by the difficulties the economy has. The draft budget for 1997 foresees the expenses for science and education at the lower level than it has been stipulated by Russian laws. If it is accepted by the Russian parliament, the scientific potential of the country will come to an irreversible distraction.

Tax laws do not encourage investments into the science and culture from non-governmental sources, that is from enterprises, companies, banks and other commercial structures.

It is a secret for nobody that the science of other states which appeared in a post Soviet space, countries of Central and Eastern Europe endure similar difficulties. Nevertheless, both in Russia and there, schools were established and struggle for their survival. They are not only a national but also an international property.

Russia is obliged to the Soros fund,Curnegi fund and other organizations for the efforts in preserving the educational, scientific, cultural potential of Russia. It is known that similar work is also shown in other countries. But it can't solve all the problems.

I am convinced that the creation of an international project aimed at preserving and increasing the scientific potential of the countries where powerful scientific potential was created and now there is a risk to lose it would have a global significance. The Congress may promote it by proposing with a proper initiative to the international organizations. The question is about the preservation of science and culture which are common to all mankind value.

3. As it was noted above, science needs increasing governmental support, including financial. Finally, it is repaid, as mankind has not such a powerful and inexhaustible source for moving towards safer and more guaranteed life, but science and technological progress.

This financial dependance is accompanied by other forms of dependence of science and scientific institutions on power which cannot be efficient. Administrative authorities, their apparatus, try to intrude upon the planning and management of science, to solve problems in this sphere which is outside their competence and needs a complete participation of the scientific community itself. Power, more seldom than ten or fifteen years ago, appeals to scientists for their advice concerning urgent and perspective problems in both domestic and foreign affairs.

Similar processes are taking place in Russia. These reveal in the above an aspiration of the state for economizing on science and education. As evidence : inactivity for more than a year of the President Council which contains the best intellectuals of Russia and should serve as a consultative board for the President when he comes to the most important, strategic decisions. That is also proved by the liquidation of the Analytical department in the President Administration. That is the department which supplied the head of the State with values, prognoses and recommendations in key questions of social and political questions and of political development of the country.

By now this process has not assumed the character of the dominating tendency. Nevertheless, the scientific community should take steps both in national and international levels to avoid deintellecting the system of state power and management, transforming science into an institution for serving the ruling circles to the prejudice of fundamental values such as the right for searching the truth and the right for criticism. Without this, mankind risks to be taken to totalitarism, to be taken unaware by hard, dramatic problems.

4. It is likely that in the twenty first century the world will enter with some disappointment in the ability of science to cope with sharp problems facing mankind. There is a deteriorated, in spite of all the efforts, ecological situation, the remaining poverty of most of the population on the planet, all kinds of social cataclysms, political collapses, wars. To a great extent in this there is one of the main, if not the main mentioned above sources - the loss of interest in science.

There is a conception that the establishment of this fact follows two practical conclusions.

The first conclusion is that it is necessary to strengthen the attention of scientists and authority to social, human consequences of scientific discoveries and their practical usage. To solve them, it is necessary to take measures in developing the humanitarian sciences. For the last 40-50 years, they have developed a lot but still they are not able to be a reliable tool in solving the problems facing society and mankind.

Now we come to the second conclusion. It is not an exclusion that the concept that had been prevailing for the last decades needs revision and more precise definition. And this concept has been linking the progress in humanitarian fields of scientific knowledge with the usage of concepts and methods of exact sciences in them.

It is not possible to argue, that in many divisions, for instance in sociology, psychology, linguistics, economical science and management, politology, law, such an approach has proved its value. But we can clearly see its weakness, insufficiency, when we come across existential problems concerning fundamentals of the human being: it's destination for the world, sense of life, relations with other people, correlation of freedom and responsibility, things common to all mankind, ethnic and personal principles, knowledge and trust, etc.

In short, everything must be done for the scientific leap that started in the twentieth century as a rapid progress mainly in natural studies and technology to continue as a qualitative breakthrough in the understanding of human nature and of society in the next century. This to my mind is the new role of science, Academies of sciences, the new duty of scientific society in the nearest and far away future.

SUBJECT INDEX

A

Academic Association 112
Academy Council 100, 101, 104, 105, 193, 195
Academy grants 27, 29, 103
Academies in transition 15, 25, 30, 31, 76
Academy of Athens 4, 42
Academies of Sciences 107, 123, 129, 131, 136, 161, 162, 165, 167–170, 216, 227, 230
  definition 161
Academy of Sciences of Albania 17, 108
Academy of Sciences of the Czech Republic 193, 194, 197
  history 194
  international collaboration 197
  science policy 195
  structure 198, 199
Academy of Sciences of Moldova 113, 115, 116, 117, 119, 123
Albania 15, 107, 108, 110
All European Academies (ALLEA) 10, 11, 15, 16, 21, 104, 105, 130, 139, 164, 169, 197
Armenia 15, 61–65, 67, 147–152, 219–222
Aryans 214
Association of University Related Research Parks (AURRP) 89
Association of University Technology Managers (AUTM) 88
Aubert J.E. 114, 116
Awards 4, 24, 89, 99, 102, 103, 141, 153

B

Balkan Academies 124, 139
Balkan countries 97, 113, 114, 117, 124, 126, 127
Balkanization 112
Baltic countries 161, 163, 165, 166, 201
Banita 212
Biomedical topics 156
Blidaru 212, 214
Bohemian Learned Society 194

Printed by Printforce, the Netherlands